Scientifically Sound

Scientifically Sound

The Science of Life

The Practice of Living

by

Peter Lacey

ISBN 978-0-578-55630-7

http://scientificallysoundbook.com

Dedicated to my courageous and loving wife, Denise, the source of all that is good in my life.

Table of Contents

Preface

One winter day, when my wife was out of town, I took my oldest son to the playground we have downtown. The playground is little more than a climbing structure and a small slide, but has the benefit of being nearby. My son, Henry, was about two and a half at the time, and that day he discovered the joy of going down the slide on his belly, his winter coat greatly reducing drag. He would slide down, I'd scoop him up at the bottom and put him back at the top, he'd slide down again. Repeat.

After half a dozen rides, something went wrong. He somehow managed to snag his face on a little piece of metal that extruded from the slide. He cried out and I gathered him up to look him over. At first, I saw nothing, but then the blood came, rivers of it, streaming from just under his right eyebrow. I threw him into the stroller, one of those modern ones with big wheels that allow the adult to run. And I ran — a good half block. My car, mind you, was sitting in my driveway, still three blocks away. I ran only half a block because I was simply incapable of running any further.

Roll that around in your head for a bit. Here I am with a small boy whose face is covered in blood and who at best is going to need some stitches. I have to go three blocks to reach my car and get him to the hospital, and I'm doing it at a *brisk walk*. My lungs couldn't deliver enough oxygen, my heart couldn't pump enough blood, and my muscles couldn't endure the strain. Until that moment, I had always been able to avoid having to overexert myself via some combination of youth, luck, or wit. But my youth was gone, my luck had run out, and nothing about that evening was even the least bit funny. When I needed to, when I really needed to, I couldn't run a single block.

As it turns out, the laceration was not serious, but it did require two stitches. A nurse wrapped my son's arms in a pillowcase to keep him from flailing, then she held him down as a doctor sewed him up. It was quick, but there was no anesthetic, and the poor frightened child screamed the scream of the damned.

When it was all over and I was pushing him back to the car, I had a smoke to calm my nerves.

When I got home I had a large dinner and too much wine. I don't remember that explicitly, but it's almost certainly so, because I always had a large dinner and too much wine. Since my wife was away, we can also assume I watched a guy movie later that evening. Probably something with Bruce Willis. Which was likely followed by a fistful of cookies and a quart of milk, before I went off to bed.

◆ ◆ ◆

You'd be forgiven for thinking that the incident just related provided the impetus for me to finally get my act together. But addiction's a bitch and life is not quite so neat. That event took place over a dozen years ago, when I was just 41 years old. In the ensuing years, I continued to fill my maw with cigarettes, sugar, and wine, all while moving nary a muscle. The results were predictable. After a lifetime of abuse and neglect, my body began to fail me pretty much constantly, not only when I needed it most. A year or so ago, the pain and shortness of breath became unendurable. I cast around for a solution, stopping by gyms and walking back out again, visiting personal trainers and balking at the price, wanting to make a change but unable to do so.

Meanwhile, my wife, Denise, while not anywhere near the mess I was, was also dealing with the limitations of weight-induced pain and constraint of motion. For a while we simply commiserated with each other, but then, to her never-ending credit, Denise actually did something about it — for her and me. She signed us up for a six-week exercise program, and that, as you'll see, really did turn my life around. It set me on the road to good health.

◆ ◆ ◆

That's what this book is about: getting healthy. It's about me getting healthy and it's about you getting healthy. Although it uses the narrative conceit of my journey from physical decline to robust health, I do not consider this book a memoir. If you are standing in the bookstore's Biography section, please show the manager the previous sentence and ask if she would consider stocking it in the Health & Fitness section instead. My personal story serves mostly as an underlying framework on which to hang the truly useful information about diet and exercise. Information you can use when it's time for you to change your life.

In these pages you will find insight and advice on how to get healthy (as well as the now-obligatory inveighing against the processed food industry). You'll find guidance on losing weight, recommendations on eating healthy, tips on how to quit smoking, reasons to add strength training to your workout schedule, discussions on the pros and cons of continuous endurance training versus high intensity interval training, and advice on simply finding the time to exercise at all. The goal, however, is not to tell you what to do, but to give you the facts you need to make the right decisions for yourself.

Moreover, my story is a proxy for your story. It's a chance for you to see a little bit of yourself in the fumbles and bumbles of an ordinary, out of shape, middle-aged man, who overcame a lifetime of bad habits and lost thirty pounds of fat, added several pounds of muscle, ended a 37-year smoking habit, ran a ten-mile race, salvaged a marriage he didn't even know was in jeopardy, and discovered the life-changing joys of being healthy.

And trust me, I truly am an ordinary man. I am not a doctor or a scientist. Neither am I a celebrity or a guru. I'm not even a professional writer. I am a software developer who works for an insurance company. I have a wife, two kids, and a dog. I live in the suburbs and drive a minivan. But by sharing a little of who I am, it is my hope that you will recognize a small piece of yourself. And by knowing I was able to save myself, you will know there's a path forward for you, too. Likely not the same path, but some path you are comfortable taking — in your own time.

◆ ◆ ◆

Possibly you're not standing in either the Biography or the Health & Fitness sections. Instead, as you leaf through these pages, you may be in the (way too small) Science & Technology section. That would make sense; fully a third of this book is comprised of biochemistry, neuroscience, and physiology. This is for a variety of reasons. For one, the science will help you better understand and navigate much of the information — sometimes misinformation or outright fraud — regarding diet and exercise that you'll find on the web, in glossy magazines, and maybe on the shelf in front of you. Put simply, a basic understanding of bio- and neurochemistry will allow you to make better decisions about your health.

While that's helpful, the real reason for all the science is because that's what health is: health *is* science. In these pages we will dig deep into the fundamental biochemistry of what makes us tick: valence electrons and the octet rule, hormones and neurotransmitters, DNA and ribosomes, galactose and chylomicrons (the two best names in biochemistry). Then bring it up a notch to answer such questions as: What exactly is polyunsaturated fat? Or a calorie? A free radical? How does a steak become muscle? Where is the weight I'm losing going? Why am I addicted to nicotine? Why am I happier now than I was 18 months ago? And, damnit, where's my runner's high?

Just as knowing about the electromagnetic spectrum and the refraction of light can lead to a deeper appreciation of rainbows, knowing how my body oxidizes food for energy and turns amino acids into proteins, keeps me exercising and eating well. I expect the same will be true for you. But if not, don't worry. I have a third reason for presenting all this science: it is simply endlessly fascinating and a great deal of fun. Just wait until you get to the bit about the ATP synthase enzyme. You'll be absolutely giddy.

◆ ◆ ◆

Now that I know the joys of being in good health, I want everyone to be healthy. But let me say right up front, I do not suffer from survivorship bias. I am fully aware that getting healthy is hard, and staying healthy is harder. There are no shortcuts, and things may not work out for you. You may need to try again and again and again. Furthermore, I have not unearthed the secret to happiness and good health, and I am not here to sell you that one sure thing. I am, in fact, deeply flawed. Hell, I still drink. A lot. It would be disingenuous for me to guarantee your good health when I'm still in the process of healing myself.

Having said that, I will point out a thread that runs through each chapter of this book — through each successive change in my life — and that is "surprise." Not just that the people who know me are surprised I ever managed to turn my life around (among my friends, the over-under for my age at time of death was 65), but that I surprised myself, over and over again. I never had any intention of getting fit or of making fitness one of my defining characteristics. I was defined by a wreath of cigarette smoke, a bottle of wine, and a package of Pepperidge Farm cookies, not by Nike Free Runners. All I ever wanted was for the pain to go away.

I had no plan to lose weight. I didn't want to quit smoking. My only goal was to get through those six weeks. The reason I set the goal posts so low was because I never thought I'd succeed at anything more ambitious. But to my surprise, I took up walking, then running. I changed my diet and lost some weight. I finally put the smokes away, and on and on and on. One damn thing after another. With each damn thing making the next damn thing a little more approachable. They led to each other almost inexorably, pulling me along until, damn, I'm fit.

So, come, take a walk with me. It's rather pleasant, and, hey, you might even surprise yourself.

Chapter 1: Decline

The sky above the estuary was painted with the stark blues and oranges of an action movie poster as my wife and I stepped onto the boardwalk. Even at 5:45 in the morning, we were not the first to arrive, a number of women had already gathered, clutching tiny barbells and nestling yoga mats. Soon, others began to appear: an older woman and her teenage granddaughter, two middle-aged women, two younger women, another woman I know from my son's school, a guy and his girlfriend, followed by even more women. I guess I expected this exercise program to be mostly female, but I admit to feeling a bit out of place, like I had entered the lady's bathroom by mistake and just made the best of it.

My wife, Denise, and I were attending our first day of boot camp, a six-week course of high-intensity exercise. The program was offered by the Fuel Training Studio, a fitness center in my north shore town of Newburyport, Massachusetts. This session, the first of 18, was being held outdoors on the boardwalk that runs briefly along the banks of the Merrimack River, just before the river broadens and empties into the Atlantic. Denise had signed us up several weeks earlier, when both of us were at the end of our tethers dealing with the discomfort that came with being 53 years old and out of shape. The hope was that a few weeks of regular exercise would put an end to the pain, and, at least for a while, allow us to function normally again.

Since this was the first time I had purposefully exercised in my entire life, I had earlier purchased all the necessary accoutrements needed to get through those six weeks. To wit: one pair of brand-new running shoes, three pairs of somewhat oversized gym shorts, two pairs of incredibly uncomfortable compression shorts that really should have been more comfortable sports underwear, three exercise shirts, two 10-pound dumbbells, and a yoga mat — we were instructed to bring those last two items.

Cutting quite the figure in my billowy silver gym shorts and unmarred white sneakers, I mimicked what the other — I guess — recruits were doing, and laid my yoga mat out on the boardwalk. Continuing to look around, I was surprised to see that everyone else was already in very good shape. Maybe a few were here to lose some winter weight, but for the most part, the group was slender and strong. In their well-fitting exercise clothes, they gathered two by two, looking more than ready to take on the world. I quailed.

There was one more guy on the boardwalk that May morning, our instructor. He spent a few minutes pacing and looking at his phone in a very work-like manner. Soon enough, though, he asked us to warm up by jogging to the end of the boardwalk and back. My newfound compatriots took off at a trot, laughing easily and discussing recent events. I groused.

The distance to the end of the boardwalk and back is maybe a quarter of a mile – call it five blocks. As you may have gathered, I couldn't jog five blocks, so I set off at an easy walk. Denise encouraged me up to a brisk walk, and by the time we completed the circuit (last), the others had positioned themselves at the end of their mats. We did the same.

The program started off with some stretches. The instructor, Tommy, described and demonstrated each stretch and everyone echoed his movements. Almost everyone, that is. I have no mind for this. Have you ever seen a movie where someone is being taught the steps to a dance, and the choreographer goes, "Step, step, dip, twirl, sweep, kick, step," or whatever, in some overly complicated way? And then the dancer repeats the whole routine back perfectly, on the first try? I always found such scenes to be hokum, thinking no one can pick things up that fast. I was wrong. As I was to learn over the next six weeks, this is something a great deal of people have absolutely no problem with. As Tommy called out various exercises and things to do with arms and legs, the cadets, including Denise, responded immediately and in unison. They looked like the Rockettes working off a hangover. I, on the other hand, cannot even keep a beat. However, by pausing awkwardly and studying the others around me, I was able to make movements largely resembling what was being requested. After a few minutes of this, we began exercising in earnest.

I obviously cannot remember each and every exercise we did that first day, but I can give you a feel for it based on my memory of the full six weeks. We did a mix of cardio things (e.g., jumping jacks) and muscle-strengthening things (e.g., push-ups). There was a fair bit or running, jumping, and skipping (honestly), as well as planks, and tricep dips, and squats, and so on.

I won't lie; it was hard. But the people from Fuel are great. Tommy knew I was an out-of-shape novice and made it clear to all of us, not just me, that we only had to work up to our abilities. He also provided alternatives and variations for exercises that might have been too difficult. As with the stretching, I simply did my best. Even though most of the class could do twice as many crunches twice as fast, I still did my crunches. If they ran, I jogged. If they jogged, I walked. Even that's not strictly true. It was more like if they ran, I ran too, until I couldn't anymore, then I walked.

When it was all over and the fit people flitted away, Denise and I gathered up our yoga mats and our tiny dumbbells and headed for the car. "You did well today," Tommy said. "Keep it up." I beamed. Then I apologized for being old and fat and out of shape. I don't know why. It seemed appropriate.

Moments later, while throwing our stuff in the back of the car, Denise asked, "Well, what did you think?"

And this next bit, people, is critical for you to understand. "It didn't suck," I answered. "I could do that again."

This was a completely unexpected response. Just 60 minutes earlier I would not have predicted this. I had spent my entire adult life performing the mental gymnastics necessary to avoid doing actual gymnastics, and now, after just one hour of exercise, I was beginning to realize just how unimaginably stupid I was.

I was a little out of breath, but nothing debilitating. I was also aware of many more muscles (and pockets of fat) than I was earlier, but I didn't hurt. Not only did I not hurt, I felt pretty good. Endorphins or something, I guessed (it wasn't endorphins; I didn't push myself hard enough to trigger an endorphin response, see Chapter 8). Not only could I do this again, I *wanted* to do this again.

We went home, where I had a cup of coffee and my first cigarette of the day. The next evening, I pulled out my phone and registered for the following day's class.

And so, for the next six weeks, Denise and I kept signing up. Every Sunday, Tuesday, and Thursday night for the next six weeks, we would have the following conversation. One of us would ask: "Did you sign up for tomorrow?" To which the other would reply with one of: "Already did," "Doing that now," "Oh, right," or "Maybe we can take the stretching class tomorrow."

Heavy, Heavy Fuel

To fully appreciate how surprising it was for me to exercise hard for six weeks, let alone enjoy it, we need to examine four more snippets in time, one for each of my principal vices. You may not be quite the train wreck I was, but I suspect you'll find at least a little of yourself in each of the iniquities that follow.

Smoking

In 1980, I was a pimply-faced, four-eyed, vice-free geek. Sadly for me, I grew up in a predominantly Irish neighborhood on the ass end of Brooklyn, New York, where vices could not be held at bay for long. In the late summer of that year, shortly before my 16th birthday, I had my first cigarette.

I have an unusually vivid memory of that evening. Along with my friend John, I was at a small house party, just a few doors up from my own. John and I were sitting across from each other at the backyard picnic table. As was mandatory even then, a red plastic Solo cup sat in front of each of us, half filled with bad beer from a pony keg. Also in front of us was someone's pack of Parliament cigarettes.

"You ever smoked?" asked John.

"Nope," I replied.

"You should try."

This was not your typical "all the cool kids are doing it" routine. John was a genuine friend, and at the time smoking was common and socially acceptable. From coast to coast, ashtrays clattered on every tabletop, cigarette butts littered the gutters, and fully one third of the U.S. population smoked. To be honest, when John offered me those Parliaments, I was eager.

I took a cigarette from the pack and put it between my lips. (In 1980, a pack of smokes cost $0.75, and it was customary to let others help themselves. Today, when cigarettes are $0.75 *each*, I imagine this is much less common.) I struck a match, applied the flame, and took two quick tokes to get the cigarette burning. With my next breath, I drew in a lungful of smoke. The aerosol — primarily tar — travelled into my lungs, irritating the until-now pristine passages. From there the smoke followed the ever-branching bronchi and bronchioles until it reached the alveoli, where, along with the intended supply of oxygen, many of the thousands of chemicals swirling around in the smoke passed into my bloodstream. A percentage of one of those chemicals, an alkaloid called nicotine, passed through the blood/brain barrier, and within seconds of that first puff, triggered a variety of chemical receptors in my brain. The effect was immediate and dramatic. There was a small jolt of pleasure and the sensation of being both relaxed and alert.

Whether you are a smoker or not, you may expect that immediately after that first drag, I was reduced to a fit of coughing. I was not. I exhaled easily, paused for a moment, and drew on the cigarette again as John looked on wide-eyed. It was easy. I loved it. I was a natural. And I immediately developed a pack-a-day habit that I maintained for 37 years, with every cigarette nudging me closer to emphysema, heart disease, and cancer.

Sloth

My lifetime of sloth doesn't have as definite a beginning as my almost-but-not-quite lifetime of smoking does, so let's hearken back to 1964. I was born into a world where, for the first time in history, it was possible to avoid physical activity for the entirety of one's recently lengthened lifetime. A world full of cars, televisions, office jobs, and people who would deliver pizza right to my door. For many of us born during or after the Baby Boom, if we didn't want to move, we didn't have to.

To most people, it was clear that a life of inactivity is not conducive to good health (though few, I imagine, knew exactly how deleterious it was). Consequently, many office workers and bank tellers embraced a lifestyle that included regular exercise — a few even included jazzercise. Some of us, however, eschewed even modest attempts to "get in shape." To us, exercise was time consuming, uncomfortable, and frequently embarrassing.

For many years, including those following that first cigarette, a distaste for exercise made little difference in my life; I was active enough. But I got married in 1998. Two years later, we moved to the suburbs and bought a house. I started commuting to work by car for the first time. In 2004, our first son, Henry, came home[1], and we moved into a larger house. In 2005 I took my first work-from-home job. In fact, from 2005 to 2013, I worked exclusively from home. In 2006, our second son, Ambrose, arrived.

During all this time, and the subsequent decade, I had little reason to move at all. I'm a software developer. My job is to poke at a keyboard all day. I enjoy this poking very much. It's a wonderful thing to enjoy what you do for a living. But if what you do every day, frequently including weekends, is sit in a chair and stare at a screen, you're going to want to set aside some time occasionally to get your butt moving. I did not set aside that time. My butt stayed motionless. Hell, my home office "desk" was no desk at all, it was a recliner! It was very intentionally

[1] Don't be confused. "Came home" is an expression adoptive parents use in place of "was born."

a recliner. I went out and bought that recliner just so I can work comfortably from home. Every morning of those work-from-home years, I'd plop myself into my recliner, pry open my laptop, and get to work, and there I would sit until the end of my day, nine or ten hours later.

If that sounds horrific, then you should know that even if you're a proud member of Team Active, if it's been more than a year since you've engaged in regular — that is, four or five days a week, regular — exercise, you're not very fit at all. After a few years of deconditioning, despite the occasional tennis match, your level of fitness has declined to much the same as mine was in 2013. Should you decide to go for a five-mile run, your lungs, while not coated with tar, are still not capable of supplying the oxygen your undeveloped muscles will need. Your muscles, in turn, are a mere shadow of what they evolved to be. Even though you are not bedridden, they have atrophied significantly since you quit exercising. The mitochondria that power your muscles are reduced in number and the capillaries that feed them have been torn down due to disuse. Your cardiac muscle is similarly underdeveloped, and is much smaller and working much harder than it needs to be. With each contraction your heart is delivering far less blood than is needed to fuel a five-mile run. Blood, by the way, that you have quantifiably less of than someone who does exercise regularly.

Gluttony

The summer of 1980 is notable for more than being the time I took up smoking: it was also when I reached my full height of six foot one. I don't remember my weight at the time, but based on belts that I still have (and can now wear again), I estimate my weight to have been about 160 pounds. If that doesn't paint a picture for you, I'll be more succinct: I was skinny. Lanky, even. In 2000, the year I moved to the suburbs and started commuting, I was still wearing those same belts. A dozen sedentary, donut-fueled years later, I weighed 210 pounds; more than 50 pounds heavier than normal and 20 pounds overweight as defined by the standard body mass index[2].

Weight gain and inactivity are inextricably linked, and those years of commuting or sitting in a recliner did not help in the slightest. That said, there is only one way to gain weight, and that is to consume more calories than you burn. Which is what I have been doing for most of the last 18 years. To be honest, my relationship with food was somewhat healthier than my relationship with cigarettes and Barcaloungers. In hindsight, I had the refined carbohydrates dialed

---

[2] https://www.rush.edu/health-wellness/quick-guides/what-is-a-healthy-weight

up way too high (so do you), but it was a pretty typical American diet. Maybe even better than most, since dinner was usually home cooked and generally wholesome.

The problem is, the typical American diet is garbage. We'll dig into this in Chapter 3, but too many of us are casually and constantly eating foods that are just this side of toxic. In particular, we are consuming far too much sugar and refined grains. In my case, that meant Multi Grain Cheerios or a donut for breakfast; a sandwich, chips, and a soda for lunch; a not unreasonable if often too-starchy dinner with wine; and then a little[3] something sweet before bed.

Three things will happen to all that sugar and starch — which, as you'll soon learn, is just sugar once removed — once digested and passed into your bloodstream. If you're active, most of it is burned up for energy, leaving carbon dioxide and water behind. A small amount is stashed away for easy access later. The rest is turned into fat and stored in your liver and adipose tissue — that is, your fat cells, a layer of tissue found under your skin and surrounding your organs. Excess sugar does not get excreted and it does not get converted into either muscle or sweat. It gets stored, and you get fat.

If that were the end of the story, all we'd have on our hands is an epidemic of pleasantly plumpness. The problem is that being overweight is making us sick — very sick. In my case, it is likely that I was prediabetic though not yet symptomatic. For others, especially those who are obese and have been for a while, it can mean full blown type 2 diabetes, coronary heart disease, stroke, atherosclerosis, metabolic syndrome, and much, much more. Associated with obesity is a list of illnesses that's every bit as deadly and expensive as those associated with smoking.

Drinking

It was only two months before that first smoke when I had my first drink. Not in someone's basement or at a house party, but on the corner of E. 37th street and Quentin Road. My friends — including John! — and I had convinced an older boy to buy us an eight-pack of nips (as we called those seven-ounce bottles of Budweiser that I don't think you can buy anymore). It wasn't difficult to get this older boy to go along. In the summer of 1980, the drinking age in New York was 18. More importantly, in 1980, no one gave a damn if some teenagers knocked backed a couple of brews. MADD wasn't to be founded for another couple of months, and nobody, least of all the police, cared very much if a grocery

---

[3] Not a little, actually. In fact, it was quite a lot. Five or six Pepperidge Farm cookies and a pint of milk.

store in a nothing part of Brooklyn was a little cavalier regarding to whom it sold beer. Still, no reason to risk anything — just find an 18-year-old, give him a couple of bucks, and he'd happily spend two minutes getting you an eight pack. After all, he was recently in the same shoes.

Beer acquired, we took it over to the corner, and made no secret at all of drinking it. It was night time, so I suppose we had the cover of darkness, but we also had the glare of streetlights. The beer was cold and, to a teenager's palate, good. No one noticed, or if they did, no one cared. We weren't boisterous or drunk, we were just hanging out talking about girls.

For the next quarter century I drank steadily, mostly in reasonable amounts. Around ten years ago, however, for no reason I can ascertain, my alcohol consumption started to creep up. Today, I drink (present tense) an entire bottle of wine every night, seven nights a week. This is not a habit that I want to continue, and it is not healthy. However, unlike every other damn thing in my life, I have not been able to change this fact[4]. In this book, my excessive drinking, the science of metabolizing alcohol, and the physical, mental, and social impact of alcoholism will be left largely unexplored. It's up to you to decide if this makes me a fraud or not, but for now, at least, I am able to function both as someone who drinks to excess and someone who is healthy and active.

Body

By the time 2012 rolled around, I was not well. I wheezed going up a flight of stairs. My neck started to hurt, then my back. I could feel my heart race just from rolling over in bed. I started wiping my ass while sitting down because I could no longer reach around while standing up (a practice I continue even now; I believe it to be the superior ass-wiping technique). I began not to like myself. I didn't like the way I looked and I didn't like the way I felt, which was like crap, throughout the day, every day.

The neck pain came and went, and predated the weight gain by a couple of years. I can vividly recall the first time it happened, in the winter of 2004-2005. It lasted for weeks. Until a day finally came when the pain simply vanished, as suddenly and inexplicably as it appeared. A year or two later, however, it would return and the cycle would repeat. By 2011 the neck pain was recurring more frequently and becoming more painful. In the daytime, these incidents were

[4] Even as this book was going to print, I quit drinking. It's been ten days so far. My goal is to be dry for one month. After which I hope to drink more casually; a bottle or two of wine per week instead of per day. That's still not healthy, but hell, I'm not a monk. If that doesn't work out, I will quit entirely.

mostly just annoying, though sometimes it was so bad that to look left or right, I'd have to literally turn my entire body. Nighttime was worse as there was nothing to distract me from the pain while I was trying to go to sleep.

As bad as the neck thing was, the back thing was worse. I didn't wake up with the back thing, it came at me in broad daylight. More like being mugged than being burgled. One second everything was normal, and then I'd do something crazy, like, say, stand up, and *SPRANG*! A flash of pain would spread all through my left (or right) lower (or upper) back, and I'd be immobilized. If I was lucky, I'd be near some ibuprofen and the acute pain would last only 20 minutes. The dull pain, however, would linger for days after. Then everything would be normal again, until the next time I reached for my phone, backed up down the driveway, sneezed, or who knows what.

Eventually, the back pain stopped going away. In 2015, it became an ever-present thing. Muted, certainly; I could get through my day. But a few things had to be approached more, let's say, gingerly. This is when the conversations started. Just a few sentences exchanged here and there between my wife and me, all pointing toward a different future.

"What are you doing?" she'd ask.

"Putting on my shoes," I'd reply.

"On the stairs?"

"I can't reach them any other way."

◆ ◆ ◆

"I need to get in shape," Denise would say. "All this traveling for work is taking its toll."

"Yeah, me too. Everything hurts all the time."

◆ ◆ ◆

"You should join a gym or something," Denise would suggest.

"I'm not really a gym kind of guy."

"You do realize you're lying on the kitchen floor?"

"Of that, I am keenly aware."

And Soul

You may be personally or intellectually aware of the physical cost of a lifestyle devoid of exercise and rich in sugar, nicotine, and alcohol. But if you're anything like me, you may not know that there's also an emotional cost associated with a hedonistic lifestyle. These habits — these addictions — are likely taking a huge toll on your family, your friends, and your ability to be happy, and you don't even know it.

What I'm referring to is the listlessness and isolation that accompanies a sedentary, over-sugared, alcohol-soaked lifestyle. This listlessness is not at all obvious, at least not to the afflicted. You are not sitting in the dark, curled up in a fetal position. Nor are you crying, depressed, or anxious. Instead, you're just kind of checked out, disengaged, lazy[5]. I never realized how withdrawn I had become until after I got healthy and suddenly I wasn't. But for the prior ten years, I was not myself. I was too quick to say no to my friends, my kids, and my wife. Each "no" seemed inconsequential at the time, but summed up to a host of lost opportunities. And I paid a price. It almost cost me dearly.

◆ ◆ ◆

I like to think I'm a "Pretty Good Dad." As a Pretty Good Dad, I've taken my two boys to amusement parks, attended the sporting events, indulged more than a few whims, and even coached Little League baseball for four years. (Technically, I was the assistant adjunct backup coach. Lacking the skills to actually coach the kids, I mostly just corralled them.) I've taken Henry and Ambrose fishing and skateboarding more often than I can count. I've tossed frisbees, thrown footballs, and kicked soccer balls.

Despite all this, the number of times I've said no to their simple requests for time with me is more than I care to think about. For every "Sure, I'll play a round of HORSE," there were three instances of "I'm sorry, not right now, son," only one of which was legitimate. Now, on occasion, "I don't feel like it" *is* a legitimate reason to not play ball with your kids. But mine were too frequent, and often because I couldn't bear the thought of chasing after a loose basketball or running down an errant frisbee. Now, Little Boy Blue is 14 and doesn't ask to shoot hoops with Dad quite as often as he used to. And Ambrose, though only 12, is already more interested in *Call of Duty* than spending time with me.

Not only was I saying no to the boy's requests for dad-time, for the same reasons, I was silently saying no to Denise's requests for husband-time. She was

[5] Incidentally, this state of listlessness used to be its own vice, called acedia, but Pope Gregory I rolled it up into sloth. Quite correctly, if you ask me.

repeatedly asking me to participate more in our marriage and family, and I, not recognizing my own faults and assuming she was just being moody, made comforting sounds and promised I would.

These discussions were sometimes tense. I know now that Denise was considering divorce, and, in retrospect, she had every right to be. Not because I was a smoker and a drinker and a layabout — to one degree or another, I'd always been those things. But now I was the guy who didn't do much of anything at all. I was becoming increasingly lethargic, shunning activities that made me uncomfortable, and actively avoiding anything that required physicality. In my ignorance, I thought I was happy enough. Looking back, however, I wasn't especially happy at all. I was just not unhappy.

Not everyone who's out of shape withdraws from the simple but uncomfortable challenges of life. But if you recognize yourself in the picture I just painted, then know that your reluctance to participate in life is not boredom and it's not a personality quirk. It is an aberrant behavior and a symptom of your declining health. More accurately, it's a symptom of addiction, even if that addiction is only to sugar.

I contend there's a straight line to be drawn between being unhealthy and being withdrawn. Just as poor health can keep you from shoveling snow, it can keep you from building strong relationships and participating in the moment. The toll of a lifetime of bad habits is not measured only in pounds and back aches, we must also count the lost opportunities to be fulfilled and content and happy. The good news, as we shall see, is that getting fit not only improves your physical health, it improves your mental health as well. But I get ahead of myself. Stick with me till the end, and you'll see just what being fit can mean to you, your family, your friends, and your engagement with the world.

To get to the end, we need to start at the beginning: finding the time.

Work to Live

When Denise and I moved from downtown Chicago to the suburbs of Boston, I began commuting regularly by car for the first time. Not to Boston, but another suburb, 45 minutes away. That job was followed by a gig actually in Boston with an absolute beast of a commute — 90 minutes each way on average, though two-hour commutes were not uncommon. It was that job that precipitated ten years of working from home. But just like you forget the torture of raising an infant and go and have another baby, I forgot the hell of that commute and took another job in Boston. This time, at least, I had the option of taking a bus, but that was small comfort.

Commutes of any meaningful distance have a negative impact on your mental and physical health, the longer the commute, the worse it is. Compared to an otherwise healthy non-commuting adult, even commutes of just ten miles have been shown to drive up a commuter's weight, cholesterol, and blood sugar, likely due to an increase in fast food consumption. Commuters are also more likely to suffer from stress, depression, anxiety, and social isolation, raising blood pressure and promoting other stress-related illnesses. With less time for family, friends, and hobbies, commuters' self-reported levels of happiness decline commensurate with the length of their commute. Obviously, all that time sitting takes a dramatic toll on cardiovascular health. In addition, commuters tend to get less sleep, stealing lost time back at the end of the day, and less quality sleep. They suffer from back aches, muscle cramps, and absenteeism. In short, commuting is killing you, every bit as much as inactivity or smoking.

So when it was time for my next job, I limited myself to either remote employers or those within a 30-minute drive from my house. I chose Liberty Mutual, a 100 year old insurance company with a beautiful headquarters in Boston's Back Bay. I could have taken an office there, but I elected to work in a boring and ugly office park that's *exactly* 30 minutes from home, with no variation in daily commuting time. In retrospect, even a 30-minute commute is 30 minutes too much. The drive has no stress-inducing congestion in it, and I tend to use the time to either catch up on the news or just be in my own head, but it still has negative consequences for my health and the environment. Moreover, the 30 minutes I spend in the car each way adds up to roughly 240 hours, i.e., ten full days, over the course of a year, ten days that I could be doing something else.

I mention all this because I am not at all sure I could have changed my life so dramatically if I had continued to spend three to four hours a day commuting. The early morning boot camp would have been right out. Even though I got up at 5:15 AM to make it there, I wouldn't be home again until 6:45. Add a shower and a halfway healthy breakfast and it's 7:30. If the commute that day is more than 90 minutes, I'm late for work. Even if my employer had flex hours, that just trades morning for night. It is possible to have a 90-minute commute and still get in shape, but it's like getting to the plate with two strikes against you.

Liberty Mutual (and likely most other established employers) also comes into play on the other side of the commute. Liberty asks for an eight-hour day. If you arrive at 9:00 AM, then leave at 5:00 PM. Arrive at 8:00, leave at 4:00. That is, there's a reasonable work/life balance, with no built-in expectation of working or being available during non-work hours. Of course, circumstances may dictate otherwise on occasion, but that's the exception and not the rule. Being healthy is not just about squeezing in some time to exercise or making better decisions about eating. It's also about having time to be with friends and family or pursuing hobbies. One way of chewing through the time that should be yours is commuting. Another, more insidious, time thief is the informal (but totally expected) 12 or 14 or 16-hour work day.

4G LTE was introduced in the U.S. in 2011; the iPhone was first released in 2007, the Blackberry in 2002. Cable modems were becoming common by 1998, and the first usable and generally available home WiFi router, Apple's Airport, became available in 1999. Prior to that, Windows 95 was the new hotness and accessing the Internet was accompanied by the sounds of small rodents squealing in pain. Maybe you're old enough to remember what work was like before smartphones and ubiquitous connectivity. If not, then know that it was glorious. Prior to 1995, when you left work, you left work. For the most part, you couldn't work outside the office even if you wanted to, and that was okay, because everyone else had left work too. This gave you plenty of time for drinks downtown, casual flirting, meaningless hookups, and all the other things that keep a young man alive.

It can be like that still; just go home at 5:00. Refuse to hook your phone up to the office email (or Slack or Skype or any other "collaboration" tool). If your boss says it's mandatory, then get your resumé up to date. There is no more social contract, they tore that up in the 80s. You are working for the money and that's all. Sure, be a team player and do the best job you're capable of doing, just do that job in eight hours, and take the other sixteen for yourself. And, dammit, don't waste four of those hours commuting.

There is something else all you young turks working for venture-backed, Silicon Valley-style startups need to learn: you need to take care of yourself. The hours you give up now are not going to be rewarded later. You are very likely not going to get the big payday. Management does not have your best interests at heart. Dinner in the office, free dry cleaning, and foosball tables are not perks; they exist to keep you at the office working. They are paying you to work eight hours, but fully expect you to turn in twice that. Rise up, comrades, you are being exploited. Software developers (and you ops guys) of the world, unite!

It could be that you are in no position to switch employers simply to get sane working hours or a better commute. Maybe you're a teacher or a nurse, and telecommuting is not even possible. Or maybe your boss is just a jerk, but there are no other employment options available. In these cases, not having gone through it myself, I can offer no advice, but you need to find some time. Can you get up an hour earlier for some light exercise? Maybe walk to the train station or get off the subway a couple of stops earlier? At the very least, start thinking about what changes you can make to achieve a reasonable work/life balance, and start working toward making them happen. One technique may be to hand your employer this book (or better still, have them buy their own copy) and point them to the next bit.

◆ ◆ ◆

Are you an employer? If so, please consider acting on what follows.

Reasonableness isn't the only reason to promote work/life balance, doing so will save you money, too. It will probably even *make* you money, and we all know how you like money.

An employee who is home by 5:30 PM (especially in the summer, when there's still hours of sunlight left) is a happy employee, and a happy employee is a healthy employee, and a healthy employee will cost you less in productivity losses and insurance premiums. Furthermore, this happy, healthy employee is going to do a better job than she would have if she were sick and resentful. In fact, the benefits of happy, healthy employees are so great you may want to further encourage them with various incentives.

And remember, we're looking for an eight-hour workday in total, we are not counting only the hours spent in the office. That means work has to stop when the office closes. You, the boss, need to discourage after-hours email, or at least discourage responding to it. Ditto all other forms of communication. Do not expect employees to link their phones to office systems. Don't make code commits at midnight the norm. Don't do anything that smacks of work during the time you are not actually paying a person to work. Be aware that if you're not going to enact such changes on your own, you may soon be forced to. There's pending legislation in New York that will make it illegal to contact employees via electronic communication after working hours.

Even if you, Mr. or Ms. Employer, limit the workday to eight hours, there's still that time lost commuting and the negative impact it's having on your employees, the environment, and society as a whole. I know you have no control over where people live, but there's still a lot of options available to help minimize the impact of commuting. For one, how about adjusting the start time to, say, 10:00? That would eliminate some of the congestion on the roads. Better still, you can allow — and by allow, I mean mandate — flex time; that is, employees should be able to work any eight hours they want. Jack can work 6:00-2:00 and Jill can work 10:00-6:00. If the work is primarily "knowledge work," then let people work from home periodically. Best of all, if the job does not absolutely require congregating in one place, like a factory or a hospital, then just mandate a fully remote workplace. That's right, get out of your lease and tell everyone to work from home. Your immediate objections, and I can hear them, are unfounded; people will not slack off, they will not be overly distracted, they will still be able to collaborate. In fact, it's safe to say productivity and efficiency will increase dramatically[6]. It's not 1995 anymore. The same technology that keeps us at work even while we're trying to have dinner can keep us at work when we're actually supposed to be working. Furthermore, your employees have just gotten weeks or months of commuting time back. So what if they spend a few minutes throwing in a load of laundry or meeting little Billy at the bus stop? Removing that small amount of stress has just made them happier still. You've also saved them the cost of gas, automobile maintenance, maybe even the cost of daycare, and you've kept one more car off the road.

10,000 Steps

My employer has not fully embraced remote work, but that's changing rapidly. They already support flex hours, and individuals may be able to work out a telecommuting arrangement with their manager. In addition, Liberty Mutual has various incentives to promote a healthy lifestyle. They will reimburse gym memberships, encourage alternative commuting strategies, and reward employees for being physically active.

On May 31st, just 12 days after my very first exercise class, I got an email from a work colleague. Liberty was running a contest: build a team of five or more people, in which each person has to complete at least 10,000 steps every business day for a month. My colleague was inviting me on to his team, which, given that he knew about the cigarettes and didn't know about boot camp, was either an enormous leap of faith or a serious lapse in judgement.

One minute earlier, I didn't even know the 10,000 Step Challenge was a thing. Two weeks prior, I would have politely declined. But at that moment, I clicked accept.

Now I needed some kind of step counter. The challenge supported three services: Garmin, Fitbit, and Apple. Long ago, I opted out of the Apple ecosystem. Had I not, I would have purchased an Apple Watch and been done with it. If you're all in on Apple, you should do just that. Me? I spent too much time researching the remaining two choices. Ultimately, I selected the Fitbit

---

[6] It's not hard to find the sources to back up these claims individually, but this page does a pretty good job of consolidating this information: https://remote.co/10-stats-about-remote-work

Charge 2. Even now, should you be in the market for a simple step counter, I can highly recommend it (or its successor), especially since it's somewhat less expensive than a full-blown smartwatch. Any of these choices, however, will count your steps, monitor your sleep, and measure your heart rate. Fitbit and Garmin devices can also last for days on a single charge. All this data is stored for you too, so you can see your progress over time.

Having pushed the magic button that would cause a fitness tracker to arrive on my doorstep two days later, I asked my wife (the owner of quite a few Fitbit trackers, the kind that clip on and which she inevitably loses) approximately how many steps there were to a mile. "Two thousand," she replied. "Why?"

"Two thousand! Really? That's five miles a day. Can I walk five miles a day?"

"What?"

I explained about the 10,000 Step Challenge.

A typical desk jockey is going to take about 5,000 steps simply going about his day. I get 100 just walking to the bathroom and back in the middle of the night. On days that corresponded with boot camp, I'd pick up around 2,000 steps from running and jumping in place, leaving only 3,000 more. On non-boot-camp days, I had to make up the whole 5,000 steps in other ways. Every time I went for a smoke, for instance, I would walk around the outside of my office building instead of congregating with the other smokers (and, yes, I see the irony, but baby steps). I started parking in the most distant spot in the lot. During my evening smokes, I'd walk around the block.

But a better solution was at hand. My wife, who's been exercising her whole life, has a set three-mile walk she will do occasionally with her girlfriends or just on her own. One weekend morning, she asked if I wanted to go. Because I was becoming the kind of person that says yes to that kind of question, I said yes.

It takes about an hour to walk three miles, and at the end of that first hour I had worked up a sweat. We walked at a pace a bit brisker than I would have used if I were alone, and I'll be honest, I found it a bit taxing. Three miles isn't really that far, but I wanted to quit closer to the two-mile mark than three. Three miles is also 6,000 steps! So I set my phone to wake me *every* morning at 5:15, not just boot camp mornings, and on Tuesdays and Thursdays I started going on three-mile walks. When with Denise, these walks would be a bit up-tempo. When alone, I would walk at a more comfortable pace. It got my heart rate up enough that it counted as exercise as far as my body's concerned.

And those extra 5,000 steps I was taking every day, they added up to burning an extra 250 calories, or to use a different scale, half a muffin

Chapter 2: Bulk

At the end of May, less than two weeks after my first exercise class and a week before beginning the 10,000 Step Challenge, I started my first diet ever. Losing weight was never part of any plan to reinvent myself, it just seemed appropriate at the time. I'm pretty sure I didn't even intend to start dieting the day I started dieting.

To understand why it seemed appropriate, we have to go back a few days. On the last Saturday of the month, Denise and I participated in a rare weekend boot camp. This was outdoors on the boardwalk like our first, and when it was done we walked over to a coffee shop that was just steps away. I ordered a croissant to go with my coffee and Denise ordered a muffin.

We took our pastries and drinks to a table outside. Several minutes later, my croissant finished, but Denise's muffin largely intact, one of the more svelte of our fellow recruits walked by. As she did, she glanced at our table and cried out, "Oh my God, you're eating a muffin!" She did this in much the same way a boxer might strike out at a sudden movement or a Clojure programmer might recoil when asked to fix some PHP code, that is, reflexively and without malice (at least in the case of the boxer).

This outcry was immediately followed by an apology, but it was too late, the damage had been done. I had been made aware that muffins are not a suitable breakfast food. From there the reasoning was clear. If muffins are not good for you, then croissants are likely not good for you either. Which probably means bagels should be shunned, and donuts thrown into the fires of Orodruin.

I exaggerate, but I'm not kidding. While it's clear there's a little too much sugar and refined flour in most pastries, I believed until that very moment that having a croissant or a danish or (for you New Yorkers) a buttered roll every morning was just fine. Not until I saw a slender and active woman shrink from a muffin like a vampire from garlic did I begin to entertain the notion that this may not be so. Why is that? What is it about a muffin that engendered such a response?

Macronutrients and Digestion

Virtually everything you eat is used to either make more of you or make what's there move. Food is used to make muscles and make muscles contract, to make brains and make brains think, to make white blood cells and make white blood cells fight disease. When it comes to powering all this construction and motion, the food you eat can be divided into two broad categories: macronutrients and micronutrients. Micronutrients are the vitamins and minerals that are absolutely essential for normal physiology, but needed in such small amounts that they are not relevant to a discussion about weight loss. Macronutrients, or just "macros," are everything else, the big things, the things you need to eat a lot of. In particular: proteins, carbohydrates, and fats. It's macros that are used for energy and growth, and for that they must be digested.

If you've prepared your food in any way, such as by cutting, curing, and especially cooking, you have kickstarted the process of digestion, the process by which food is broken down into its base components for use by the body. Cooking in particular has been likened to a second stomach, using heat to do much the same work your digestive system would otherwise have to do, and thus allowing easier access to all those calories. The actual biological process of digestion begins in the mouth, where chewing and saliva start breaking down food and preparing it for a trip down the alimentary canal. Saliva is made up of a great deal of substances. It's mostly water, obviously, but among other things, it also contains enzymes, notably amylase and lipase, which begin to break down starches into simpler sugars and fats into fatty acids, respectively.

That last bit raises a number of questions: What are enzymes? What are starches? And what are fatty acids? We'll save the latter two for later in this chapter, right now we need only a cursory understanding of enzymes. A more complete account will be given in Chapter 6.

Most enzymes are proteins (we'll discuss proteins in depth in Chapter 5). An enzyme acts as a catalyst for chemical reactions in the body. Found in your cells, enzymes speed up the process of stitching molecules together or breaking them apart. That is, an enzyme accelerates the conversion of one molecule, such as the polysaccharide starch, into other molecules, such as the monosaccharide glucose. Because enzymes are highly specialized, each catalyzing just one type of chemical reaction, there are tens of thousands of different types in the human body.

As far as digestion is concerned, we are interested in just three broad classes of enzymes; those that break down carbs, those that break down proteins, and those that break down fats. Your body can't use the food you consume in anything close to its whole state; it must metabolize it down into much simpler molecules that are then passed into the bloodstream and put to use. Everything you eat that can be broken down is broken down: carbohydrates into monosaccharides, proteins into amino acids, and fats into fatty acids. But before we can take a look at each of these, we need to brush up on our chemistry.

Chemistry 101

An atom has a nucleus made up of one or more protons. The nucleus may also contain one or more neutrons, but having no charge, neutrons are chemically uninteresting and will be mentioned no more. Orbiting around the nucleus are one or more electrons. In its normal state, an atom has one electron for every proton. Hydrogen, the simplest element, has one proton and therefore one electron. Helium, the next element over on the periodic table, has two protons and two electrons. Lithium has three of each, beryllium four, and so on down the line.

If you care to, you can picture these electrons making stately two-dimensional circular orbits around the nucleus, much like planets orbiting a star. That would be wrong, of course, but it's easier for our monkey-brains to handle than the reality of electrons zipping around and about the nucleus in a cloud of probabilities and strangely shaped orbitals. Electron orbits differ from planets in another way: they are always at a discrete distance from the nucleus. Each of these fixed orbits, these "shells," can house a set number of electrons before they are full. The innermost shell has room for two electrons, the next has room for eight, then eighteen, and so on, where "and so on" equals 2 × the square of the shell number.

Now here's the good bit. For certain elements — a collection called the "main group", which includes all of the biologically interesting elements: carbon, nitrogen, etc. — no matter how many electrons they are *supposed* to have in their outermost shell, formally called the "valence shell," they really *want* to have eight. When a main group element has eight electrons in its valence shell it is stable and unreactive. This tendency is commonly referred to as "the octet rule." Should an element not naturally be in accordance with the octet rule (that is, not a "noble gas," such as neon or argon), there are a number of ways it can get there. If the element has just one or two electrons in its valence shell (valence electrons), such as with sodium or potassium, the element can give them up to some other needy atom. This means the element will no longer have an equal number of protons and electrons. Such an element is known as an "ion," and is said to be "ionized." Since protons are positively charged and electrons negatively charged[7], an element that loses one or more electrons is also positively charged. Conversely, elements with nearly full shells may accept some other element's electrons, forming a negatively charged ion. When these two things happen simultaneously, as they mostly do, with one element shedding an electron and another picking it up, then these two, now oppositely charged and thus attractive elements will form an "ionic bond." For instance, sodium may lose its sole valence electron, while chlorine, with seven valence electrons, picks it up. The two oppositely charged ions now attract each other, bind together, and form a stable molecule of sodium chloride, better known as table salt[8].

One ion is of particular interest to us, the positively charged hydrogen atom, denoted H^+. H^+ is a hydrogen atom that lost its electron. In other words, it's just a proton. Which you would think would be no big deal, but it's the ability of a molecule to give up a hydrogen ion that makes it an acid, and an excess of protons in a solution makes it acidic. Acids and acidity come up a lot in biology. In fact, we've already encountered two of them: fatty acids and amino acids.

Another way for elements to fill up their valence shells is to share electrons, thus forming a "covalent bond." Covalent bonds are more common than ionic bonds. When two or more elements share electrons, the electron is counted twice in regards to the octet rule, once for each element's valence shell. For example, oxygen has eight protons and, thus, eight electrons. Two of those electrons are happily orbiting in the innermost shell, leaving six valence electrons and room for two more. One way to fill those gaps is for the oxygen atom to share two electrons (a double bond) with another oxygen atom. The resulting diatomic molecule, O_2, aka oxygen gas, is the common form of oxygen in the air we breathe. Alternately, oxygen can fill those two gaps by sharing the single valence electron of two hydrogen atoms, while sharing two of its electrons right back, thus making H_2O. There are many other possible configurations, of course, and not all of them result in highly stable molecules (e.g. ozone, O_3).

The interaction of elemental electrons is a large part of what we know as chemistry, and there are more ways for elements to interact than just ionic and covalent bonding. Hydrogen bonds, for instance, which are responsible for collecting a bunch of H_2O molecules into water, and for connecting two strands

---

[7] Which does not mean *without* charge, but rather of *opposite* charge.

[8] Some ions, the electrolytes, are physiologically interesting in that they can make a solution electrically conductive. The movement of electrolytes in and out of cells is critical to the proper regulation of electrical communication in muscles and neurons.

of DNA together to make a double helix, are hugely important. But it's covalent bonds that are of primary interest to us, as they form the building blocks of organic (i.e. carbon-based) molecules, especially as pertains to bonding with hydrogen, nitrogen, and oxygen atoms. It's some combination of these four elements that make up the bulk of the biosphere. Carbon is of particular interest, because it has four valence electrons, and that means it's not likely to shed them nor find some element that can donate four more. This makes carbon good husband material; it likes to form stable covalent bonds with other elements, which is why life is carbon-based[9] and not, say, based on fluorine, with its seven nasty, highly reactive, unstable valence electrons.

With Chemistry 101 behind us, we can turn our attention once more to the macronutrients we consume: carbohydrates, protein, and fat.

Carbohydrates

Carbohydrates are molecules that have one or more carbon atoms (that's the "carbo" part) covalently bonded to one or more water molecules (that's the "hydrate" part). There are a small handful of molecules fitting this description that are not considered carbohydrates (such as formaldehyde). Therefore, carbohydrates, as commonly discussed, have at least three carbons and three water molecules. Carbohydrates are also known as saccharides, that is, sugars. There are four classifications of saccharides dependent on how complex the molecule is. The simplest are the monosaccharides, the fundamental unit of carbohydratitude (an *extremely* recent coinage).

Monosaccharides cannot be broken down any further and still be considered a carbohydrate. Glucose (known as dextrose when it's not floating around in your body) and fructose are the most common monosaccharides. They are chemically similar in that they both have six carbons, twelve hydrogens, and six oxygens ($C_6H_{12}O_6$), but these base elements are arranged in subtly different ways.

If you take two monosaccharides and stitch them together, you get a disaccharide. The most common disaccharides are sucrose (table sugar), lactose (one of the few carbohydrates not derived directly from plants), and maltose (found in malt, i.e., germinated grains). Together, the mono- and disaccharides are commonly, but unscientifically, called the simple sugars.

---

[9] And why science fiction authors frequently posit silicon-based lifeforms, since silicon, immediately below carbon in the table of elements, also has four valence electrons.

If you join between three and roughly ten monosaccharides together, you get the oligosaccharides. Some of these mildly complex sugars are digestible; others are not, since humans lack the enzymes to break them down. Indigestible oligosaccharides, sometimes known as prebiotics, get passed into the small intestine and on to the large intestine, the colon, where gut microbiota may be able to eat them, releasing a small amount of useful energy. While the microbiome plays an important role in overall health, and even in digestion and nutrient extraction, the subject goes beyond the scope of this book and won't be discussed further.

If you take a larger bunch of monosaccharides and bind them together, you get the polysaccharides, most notably starch, glycogen, and cellulose. All of these polysaccharides are comprised solely of glucose, but the glucose is chained together in different ways; that is, they are glucose polymers[10]. The first two substances, starch and glycogen, serve the same purpose, acting as a reserve form of energy for when glucose monosaccharides (glucose monomers) are scarce. By and large, starch is made by plants and glycogen by animals. Cellulose is different; it's used to form the cellular walls of plants and is a type of fiber.

Let's say you're noshing on a peach. A medium-sized peach weighs roughly 150 grams, most of which is water. Most of what's left is carbohydrates, but there's also a trifling amount of fat, protein, and micronutrients. As you chew, you expose more and more of the peach to the amylase enzymes in your saliva, which get busy breaking down the small amount of starch found in a peach. When the bolus that was a chunk of peach hits your stomach, cells that line the stomach secrete, among many other things, hydrochloric acid (HCl). Not only does the HCl attempt to kill any pathogens that may have hitched a ride, but it neutralizes the amylase and, along with the stomach muscles, turns the peach into a slurry called chyme. The highly acidic chyme is then slowly released into the duodenum, the beginning of the small intestine, where the magic begins.

About five grams of a peach are either fructose or glucose. Being monosaccharides, these are passed through to the lower intestine without further processing. However, there's roughly eight grams of sucrose in a peach too, and that needs to be broken down before it can be absorbed. Sucrose, a disaccharide, is made up of a glucose molecule joined to a fructose molecule. To bust sucrose down into its constituent components, cells lining the duodenum release the enzyme sucrase (all of your better enzymes have names that end in either "ase"

---

[10] The suffix 'mer' means part, segment, or unit. Combined with a prefix, it tells you something about the structure of a molecule. For instance, 'iso' means 'same' and thus glucose and fructose are isomers because they are made up of the same parts. The prefix 'mono' means 'one', so a monomer is the base unit under discussion. 'Di' means two, so a dimer has two of the same parts. And so on, up to 'poly' meaning 'many;' a polymer is made up of many of the same parts.

or "sin," and tend to begin with a reference to the substrate they catalyze). The sucrase enzyme binds to sucrose, and only sucrose, and snips it in two. The resulting monosaccharides now get to continue the ride into the jejunum, stage two of the small intestine. (Note, this metabolic snipping would have happened anyway, since the bond would have simply dissolved in the stomach acid. The enzyme is there just to speed things up, as enzymes do.) Once in the jejunum, the glucose is transported into the body proper via specialized epithelial cells, called enterocytes, that line the intestinal walls. The fructose takes a slightly more arduous journey through the intestinal wall, but the process is much the same.

While there are a few other carbohydrates in a peach, all except one are in quantities too small to care about. The one remaining carb of note is cellulose, and there's about 2.5 grams of it in a peach. Cellulose is a polysaccharide fiber composed of thousands of glucose molecules. The human body produces no enzymes that can break down cellulose, so it just continues through the intestines. Ultimately, while the monosaccharides are absorbed through the enterocytes, the fiber will make it all the way through the alimentary canal and get pooped out the other side. Which is odd given that TV commercials and cereal packages have conclusively shown that fiber is an important part of a person's diet. How is that so if it just passes right through you? Let's pause briefly to clear that up.

Since fiber isn't processed by the human body, it has no nutritional value. However, it has an important role to play in regulating digestion and bowel movements. For one thing, it shields the simpler carbs from their enzymatic butchers and thereby keeps all that sugar from hitting your bloodstream at once; that is, fiber-rich foods have a lower glycemic index[11]. In fact, fiber keeps a fair amount of sugar from being metabolized at all, instead, sending it on to feed the bacteria in your gut. This is because whole fruits and vegetables have two kinds of fiber: soluble (e.g. pectin and inulin) and insoluble (e.g. cellulose and lignin). In the stomach, these fiber types play off one another, with the soluble fiber forming a gel in the gut that slows digestion, causing sugar and other nutrients to be absorbed more slowly, while the insoluble fiber adds bulk to your stool and enables food to pass more quickly through the stomach and intestines.

Fiber also reduces body cholesterol, though no one's exactly sure why. Most likely it's due to fiber absorbing bile in the intestines, preventing it from being reabsorbed and forcing the liver to make more. Since bile contains cholesterol, that cholesterol is no longer free to (debatably, stay tuned) gum up the works. Finally, and importantly to those looking to lose weight, fiber makes you feel full.

---

[11] The glycemic index is a value between 0 and 100 assigned to foods based on how quickly those foods raise glucose levels in the blood. Foods with a low glycemic index release glucose slowly. Foods with a high glycemic index release glucose rapidly.

Think of the difference between eating an orange and drinking the equivalent of six oranges in a glass of OJ.

Protein

Let's leave what's left of our peach in the jejunum for a bit, and assume that along with that peach you're also having, oh, a well-marbled steak, a good source of our other two macros: protein and fat. Protein first.

Most of the important parts of you are made out of proteins — your brain, your muscles, your hormones, pretty much everything. These proteins are themselves made up of amino acids. If you are at all concerned about a healthy diet, you will hear a lot about amino acids, but it's not sufficient to say "amino acids are the building blocks of proteins" and leave it at that. That's deeply unsatisfying. What are they? What makes an amino acid an amino acid? At the same time, diving too deep into the chemistry of amino acids is not particularly helpful, even for the interested layperson. Thus, we must tread the fine line between too much and too little information.

Amino acids are comprised of three distinct components: a collection of atoms known as an "amine" group, another collection known as a "carboxyl" group, and a final collection of assorted atoms known as the "side chain." In organic chemistry, the term "group" has a very definite meaning. It refers to molecules — collections of chemically bonded atoms — that recur commonly and react with other chemical groups in well-known ways. These groups are generally found attached to a carbon backbone along with one or more other groups. Finally, these chemical groups will have a collection of other atoms attached, known as "the rest," or just "R," that give the molecule its unique properties. In other words, a chemical group allows us to ignore the individual atoms and focus on the consistent interactions of larger molecules.

An amine group is a nitrogen atom (five valence electrons) that is bonded to three other atoms. (If those three other atoms were all hydrogens, we'd have ammonia. That's where the name comes from, though I'm not sure what happened to the missing "m.") In an amino acid, those three other atoms happen to be two hydrogens and a carbon. A carboxyl group is slightly more complicated: it has a central carbon that's double bonded to an oxygen, single bonded to an oxygen-hydrogen (hydroxyl) group, and single bonded to one other atom or molecule — its R. Chemically, that's R-COOH. It's that hydrogen atom at the end, by the way, that tends to get ionized and dissociate from the rest of the carboxyl group, making the molecule an acid — carboxylic acid, to be precise. It's the R on the left, however, that's of interest to us, because if the R happens to be a carbon atom that's also connected to an amine group, then we have the backbone of an amino acid. And if the carbon atom attached to the carboxyl group is the *same* one that's attached to the amine group, we have a special type of amino acid called an α (alpha) amino acid. These are special because it's α-amino acids that get built up into proteins that get built up into you. That is, they are coded for in our DNA.

Also attached to that central carbon atom is the side chain. Since every amino acid has amine and carboxyl groups, it's the side chain that differentiates one from the other. The side chain may be as simple as a single hydrogen atom (glycine) or it may be a complex arrangement of carbons, hydrogens, oxygens, nitrogens, and even sulfur.

Not a whit of all that chemistry is useful if you just want to lose weight or build muscle, but questions need answers. In Chapter 5 we'll learn how your cells take the protein-building instructions tucked away in their nuclei and use them to construct proteins from these amino acids. For now, know that the amino acids are joined together by peptide bonds, which are formed when the carboxyl group of one amino acid covalently bonds with the amine group of another, releasing water and making a dipeptide. Similar to how saccharides are named, adding more amino acids gives you oligopeptides and polypeptides. If a polypeptide is long enough, around 50 amino acids or more, we call it a protein. Proteins bend and twist and fold into a variety of shapes, and it's these shapes that largely determine their function. How proteins fold is determined by which amino acids are used to build them and in which order. In humans, just 21 amino acids[12] are used to make all of the different proteins in the human body. Of these, the body is capable of synthesizing twelve of them; the other nine — the so-called "essential" amino acids — must be consumed as food.

Which brings us back to that steak. After it's been cooked and masticated (starting the digestion process), it finds its way into your stomach and a bath of hydrochloric acid and sodium chloride. This gastric acid is used to "denature" the proteins; that is, to unfold them and expose the peptide bonds. The stomach also releases an enzyme called pepsin, a protease which, based on the name alone, we can assume is involved in breaking these newly exposed peptide bonds. Pepsin goes to work on the bonds between certain types of amino acids, snipping the protein into smaller and smaller chains. Additional proteases lie in wait in the duodenum, where they will break down the remaining polypeptides until nothing is left but free-floating amino acids, which, like the monosaccharides before them,

[12] There are about 500 known amino acids, but only 22 of them can be stitched together to form a protein and one of those is not found in humans.

are passed into the jejunum for absorption into the body. We'll leave them there with what used to be a peach.

Fat

Reach down and grab a handful of skin around your waist. Feel that? If you're like most Americans, you're holding on to a fair bit of fat. That fat, it's important to note, is not solely the fat you've consumed. The fat you have eaten is very likely not even a significant percentage of the fat around your waist and in your thighs and butt and seemingly everywhere else. But before the big reveal, let's find out what fat actually is.

Just as proteins are made of amino acids, which are in part built up from carboxylic acid, fats are made up of fatty acids, which also contain carboxyl groups (hence the 'acid'), but with a long chain of carbon atoms hooked onto them, like the tail of a kite. Carbon, remember, has room in its outer electron shell to bind with up to four other elements. In a fatty acid, two of those are the neighboring carbons in the chain, the ones above and below it. The other two, left and right, are hydrogens. In this configuration the fatty acid is said to be "saturated," that is, all the hydrogen that can be present is present.

However, it's possible for two carbon atoms to share *two* of their electrons, forming a double bond. If a fatty acid has a double bond connecting neighboring carbons, then there's no room for one of the hydrogens that would otherwise be there, and that fatty acid is said to be "unsaturated." If there is more than one double bond in the chain of carbons, then that fatty acid is "polyunsaturated."

In addition, you can classify unsaturated fats from where the first (starting from the non-carboxyl end — i.e., the bottom) double bond is encountered. For instance, if it's in the third position, then that fat is an "omega-3." If it's in the sixth position, it's an "omega-6." Now, take three strands of fatty acids, hook them onto a glycerol molecule like ties on a rack, and you have what we commonly call fat, and less commonly call triglycerides. For the record, a glycerol molecule consists of three hydroxyl (OH) groups linked together via a three-carbon backbone.

Saturated fats, where all the carbons in the tail are bound to two hydrogens, are physically straight and can line up next to each other in neat little rows. It's because of this that saturated fats, as found in meat and dairy, are solid at room temperature. In unsaturated fats, though, at least two carbons, the double-bonded ones, hold on to just one hydrogen each. If both those hydrogens are on the same side of the chain (the so-called "cis" isomer), as they almost always are in the natural world, it will cause the chain to bend a little. The more double bonds like this there are, the more pronounced the bending. Therefore, unsaturated fats, as found mostly in plants and seafood, are liquid, that is, oil, at room temperature. Both saturated and unsaturated fats are an important part of your diet and are required for good health. Mostly these fats, like carbohydrates, are used for energy production. But fat is also essential for processing certain vitamins, regulating hormonal activity and inflammatory response, preventing certain diseases, and buffering vital organs against shock. Fats are also the main component of cell membranes.

Because they're so much in the news, I do need to discuss trans fats. For the most part, you won't find a lot of trans fats in nature. There are a few in milk and in meat, but they're generally only about 5% of the entire fat content. The trans fats you more commonly come into contact with are man-made, and also go by the name "partially hydrogenated vegetable oil." By now, you should be able to make a guess as to what "partially hydrogenated" really means. Take some unsaturated fat, that is, vegetable oil, break the carbon double bonds (with heat), inject some hydrogen to fill up some but not all of the resulting gaps, and you have partially hydrogenated vegetable oil. (If you filled all the gaps you would have a fully hydrogenated, i.e., saturated, vegetable oil, otherwise known as Crisco.) Trans fats, so called because the hydrogen atom ends up on the other side of the unsaturated carbon chain from the more typical cis fats, have a number of useful properties — if you are in the processed or fast food industries. They are solid at room temperature, yet melt easily; they are inexpensive to produce; they don't spoil as readily; and, of course, being fat, they taste good. The problem is, for reasons not fully understood, trans fats seem to play a substantial role in promoting heart disease and stroke. We'll come back to the relationship among fat, cholesterol, and cardiovascular disease shortly. But right now, let's catch up with our steak.

How much fat is in a steak depends greatly on the cut. Since we want to enjoy our imaginary steak, we made sure it's a ribeye, the fattiest cut. Fully half the weight of a ribeye is fat, glorious fat. A 12-ounce (340 gram) ribeye has 35 grams of unsaturated fat and 22 grams of saturated fat (as well as 90 grams of protein and about 430 mg of cholesterol). Fat digestion begins in the mouth, partly by mastication and partly due to the actions of the salivary lipase enzyme. After swallowing, fat escapes the stomach without further malice. Sure, it's been mashed into chyme along with everything else, but it is unmolested by any gastric enzymes or other secretions.

It's when the chyme gets to the crucible of the duodenum, however, that things start getting hot for triglycerides. To understand why, you need to know that one of the many functions of the liver is to make bile (out of cholesterol, as earlier noted, among many other things). Bile is mostly water, but it also contains "bile salts," which act as a surfactant that can wrap up water-loathing globs of fat and keep them from glomming together in an aqueous environment. That is, bile is pretty much dishwashing soap. The bile the liver makes is transferred to the gallbladder for safekeeping. This is, by the way, the only job the gallbladder has; it's just a bag of concentrated bile (for this reason the gallbladder is sometimes referred to as the Hannity Sack).

When you eat, the gallbladder squirts bile into the small intestines. The duodenum, like the stomach, expands and contracts, breaking apart the globules of fats. These fat droplets get surrounded by the bile and kept in suspension (emulsified). This greatly increases the water-exposed surface area of the fat, which is good because next up is pancreatic lipase. The lipases excel at the close-up work, cutting the triglycerides into two free-floating fatty acids and one that's still attached to the glycerol molecule. These are all ultimately shipped down the intestines and absorbed into the body through the enterocytes.

At this point our meal of steak and peach has been digested, broken down into its base nutrients, and passed through to the far side of the intestines. Let's leave it there for a bit. In later chapters we'll discuss how these macros are used for protein synthesis (Chapter 5) and cellular energy (Chapter 6). In this chapter, I wanted to primarily cover the main macronutrients and the digestive system. But since this chapter is also about weight gain, we need to discuss why we're all so fat in the first place. That discussion will benefit greatly from a crisp understanding of calories.

What Exactly is a Calorie?

What does it mean for a chicken breast to have 300 calories? How can something as different as a chicken breast and a muffin be reduced to the single concept of a calorie? Are chicken calories the same as muffin calories?

A calorie is a unit of measure, similar to "temperature" or "speed." A calorie is not a substance. You cannot hold a calorie in your hand any more than you can hold a Fahrenheit. Unlike temperature and speed, a calorie is a measure of "energy," that is, of useful work. In this way, a calorie is more akin to a BTU, joule, or electronvolt. Precisely, a calorie is equal to the amount of heat necessary to raise the temperature of one gram of water by one degree Celsius, from a temperature of 14.5 degrees Celsius, at a pressure of one atmosphere. A definition which, while technically correct, is not terribly useful. Especially since what we commonly call a calorie is really 1,000 calories, a kilocalorie[13].

Even though everything that can burn contains calories[14], the calorie unit of measure is typically reserved for specifying the energy content of food, and food is not considered to have any energy unless it can be metabolized by our cells. Thus, sugar, fat, protein, and alcohol all have calories, but water and (insoluble) fiber do not. You can determine the calorie content of a substance by putting it in a container, placing the container in water, completely burning the substance, and measuring the temperature change in the water. Or you can rely on the fact that someone's already done that and determined that on average carbs and proteins have four calories per gram, fat has nine, and alcohol seven. If you happen to know the macronutrient count of a substance, you can just do some simple math to get the calorie count.

Your body needs energy to stay alive, deriving that energy from the foods you eat. It needs more or less energy depending on a wide variety of factors: sex, age, activity level, etcetera. If a particular individual needs 2,000 calories a day to get by, that person needs to consume at least 2,000 calories worth of macronutrients. If they don't, they will begin to lose weight. Eventually, this person's metabolism will start to slow down to match caloric intake. If the calorie deficit persists, cells will start to die and so too will the individual.

In this way, it's easy to assume that when it comes to weight loss, a calorie is a calorie is a calorie. This is not wrong; you will not lose weight unless you are consuming fewer calories than you burn. However, this concept of calorie equivalency is far too reductionist. In reality, protein is different from fat is different from carbohydrates. Similarly, complex carbs are different from simple carbs. Even people are different from other people; I may extract more calories from a chicken leg than you do. And both of us extract more calories in the morning than we do at night.

Similarly, calories differ in what, if any, hormonal response they trigger. For example, glucose triggers an insulin response, but fructose does not (at least, not immediately), even though these two molecules are nearly identical. Because the body responds differently to fat than sugar, a diet high in fat will keep you feeling full for much longer than a diet high in carbs, something that's very important when it comes to weight loss.

---

[13] Convention has it that when referring to kilocalories, kcal, one should capitalize the 'C.' But convention is not very conventional, and I, like everyone else, am going to ignore this rule and use the lowercase. Just know when I use the word "calorie" from here on in, I am referring to kilocalories.

[14] A gallon of gasoline has about 31,000 calories. As does a gallon of vegetable oil. Gasoline just releases its energy faster. Dramatically faster.

Calories wrapped in fiber are different from bare calories, since the fiber will slow down the rate at which food is absorbed into the bloodstream. As previously discussed, it may keep some food from being processed at all. How fast carbohydrates hit your bloodstream matters greatly since foods with a high glycemic index trigger insulin spikes, and the spikes trigger more aggressive fat storage by adipose tissue and are linked to eventual insulin resistance. This is one of the many reasons why an apple is infinitely better for you than apple juice. The fiber itself, while having the same four calories per gram as other carbohydrates, is not fully digested, and thus virtually calorie free.

Moreover, it's important to realize that it takes energy to extract energy, and the various macros differ in how easy they are to metabolize. It takes around 25 calories to break down and consume 100 calories of protein, but only about 3 calories to metabolize 100 calories of fat. Furthermore, it takes less energy to process food if it's been cooked, so you'll get more calories from cooked broccoli then you will from the same amount of raw broccoli. It's even the case that starchy leftovers have fewer calories than before you put the casserole in the fridge for the night.

While this book doesn't cover vitamins and minerals much, for good health your body absolutely needs those micronutrients found in whole, natural foods. You cannot live on a diet of candy bars, soda, and junk food, regardless of the calorie count.

Saying that calories from sugar and starch is the same as calories from chicken is like saying BTUs from coal are the same as BTUs from wind turbines. Yes, both coal and wind produce energy that can be measured on the same scale, but one is quantifiably better for the environment than the other. While it remains true that you will not lose weight if you consume more calories than you burn, calories are a very crude measure. It's more important that you eat the right kind of calories, than a certain number of them. For your own sanity, please do not start obsessing over the caloric content of everything you eat, that way lies madness. Besides, most calorie counts are wrong, anyway.

Why Are We Overweight?

Broadly, the macros we consume either make more of us (proteins) or are burned up for energy (fats and carbs). We become overweight, very simply, when we eat more than we need to fulfill those two needs. A more nuanced question than "why are we overweight?" is why are so many Americans (and, increasingly, the rest of the world) overweight when we know more and more about the negative consequences of obesity? After all, smoking has seen a dramatic decline over the same period that our waistlines have expanded. Why is eating different? There is no single correct answer. We are fat for many and various reasons, not all of which apply to every overweight person. It's even posited that one of the culprits could be the decline in smoking itself, since people who quit smoking tend to put on weight. While the contribution of ex-smokers to America's obesity epidemic is likely negligible[15], the analogy to smoking is still illustrative.

It's not easy, but when it's time to quit smoking, you don't have another cigarette. And when it's time to quit drinking, you don't have another drink. But when it's time to lose weight, you can't just quit eating. You can try to quit excess sugar and fat, but every time you bite into something, you're awakening that part of your brain that gets a jolt of pleasure from eating, and your resolve will start to fade. Eating is pleasurable, especially if what you're eating is fatty or sweet, and people like to do pleasurable things. Indeed, we sometimes find it hard to stop.

The packaged and prepared food industries are fully aware of this, and happy to deliver their carefully crafted bounty unto us. The sugar and fat content, the salt and MSG, the texture, the color, the smell, the packaging, the presentation, every last aspect of nearly everything you eat is orchestrated to maximize your pleasure, often at the expense of your health. In Chapter 4, we will learn how the reward system in your brain is responsible for addiction. That information will be presented in the context of smoking, but every word of it applies to food, as well. That is, it's possible to become addicted to food. More precisely, it is possible to become addicted to highly palatable food: sugar, fat, salt, etc. Since addictions need to be satisfied, we eat, and then we eat some more. There are many underlying biological, psychological, and sociological factors involved in food addiction, but that there's something to get addicted to in the first place is what's at issue. No one ever got addicted to broccoli or eggs or even the occasional sweet. It's the seemingly infinite amount of high-calorie junk food that fuels food addiction.

Relatively new to the human condition is that there actually is a seemingly infinite amount of calories. Right up until the middle of the 20th century, it was difficult for most people to get fat even if they were metabolically or psychologically predisposed to being so. Even when there was plenty of food, it was mostly "real" food: meat and fish, fruits and vegetables, milk and eggs, and bread from the baker. The post-war caloric explosion that altered America's diet is primarily fueled by processed food rich in sugar, starch, fat, salt, emulsifiers, stabilizers, thickeners, and preservatives. These calories are inexpensive, convenient, tasty, and pretty much everywhere. Whether it be the Hostess

[15] Many smokers become ex-smokers by the simple expedient of dying.

cupcake at the gas station or the burger, fries, and Coke from McDonalds or the frozen pizza from the supermarket, cheap, delicious, and empty calories are always at hand[16].

While toothsome, these products are very far from wholesome. High in sugar and low in nutrients, junk food has displaced the calories we once got from real food. And don't imagine that the list of processed food is limited to the obvious culprits of potato chips and soda. It also includes just about everything in the frozen food aisle of your local supermarket, as well as most of the bread aisle, the pasta aisle, the snack aisle, the drinks aisle. Basically, all the aisles. To simplify, nearly every edible product for sale in a supermarket or fast food joint has too many calories and too few nutrients. Nor is the fast food moniker limited to the likes of McDonalds and Burger King; it also includes Starbucks, Applebee's, Denny's and virtually every other chain restaurant.

That our markets and restaurants are brimming with dangerously addictive calories is only half the problem. The other half is that we're barely moving as we eat them. In the previous chapter we discussed how difficult it can be to find time to exercise when not at work, and about the negative consequences of commuting. But what about our activity levels while at work? The last fifty years have seen a dramatic decline in the number of farmers and fishermen and factory workers, and a corresponding rise in the number of telemarketers, software developers, and service workers. In 1960, 48% of all jobs in the U.S. required at least moderate levels of physical activity; today that number is just 20%. Measured in calories, American men are burning 140 fewer calories per day than they were 50 years ago, and women 124[17].

In Chapter 1, we placed a lot of the onus for long commutes on employers, but one of the reasons commutes got longer is that people (white people, that is) actively chose to live further away from the office, abandoning the cities and moving to poorly planned suburbs. Suburbs without sidewalks, without shops, and without parks, forcing everyone into the car for even the most mundane tasks. You couldn't walk to the corner store even if you wanted to, because there was no corner store, or even a corner. As the commutes got longer and free time became more scarce, those convenient if unhealthy calories got even more tempting. Why shop and cook when you could pick up a bucket at KFC?

These days, what little free time we afford ourselves is increasingly consumed by our electronic devices. In 1960, nine out of ten houses had a television — usually just one — that aired programming from at most three networks, and

[16] It is indisputably a good thing that most of the world no longer knows famine, but the same results can be obtained without feeding the world junk food.

[17] https://journals.plos.org/plosone/article?id=10.1371/journal.pone.0019657

then only during waking hours. Don't get me wrong, *Bonanza* and *The Twilight Zone* made for compelling viewing, but even if you add in an occasional movie at the theater, it's small potatoes compared to the entertainment options we have today. Among broadcast, cable, and streaming options, there is essentially an unlimited amount of high (and low) quality entertainment. Billions of hours of distraction are available all day, every day, in every room of the house.

Or wherever else you may be currently standing — since each of us is carrying around a supercomputer in our back pocket. Not only are our mobile devices capable of streaming just about all available content (except *Star Trek: Discovery*, which is only available on CBS All Access, and, honestly, no one's going to subscribe to that), it's also capable of stealing all of your remaining attention and delivering it to Facebook, Twitter, and Reddit. In 2017, the average adult spent nearly 24 hours a week online, 17.6 of them while at home[18]. Fifteen years ago, this was barely even possible.

Despite all that sitting around, self-reported levels of physical activity haven't changed all that much in the past 30 years, even as our belt sizes have. Certainly, the kind of calories we eat plays a part in that, but so too does the sheer number of them. Starting in the 1970s, portion sizes have increased steadily in the U.S. and around the world. Not only is this evident in the king-sized Milky Way bar (480 calories) and the 128-ounce Team Gulp soda from 7-Eleven (1,480 calories), it's famously visible in the typical fast food menu of Baconators (980 calories) and Dave's Triple Combos (1,740 calories). But you don't need to eat packaged or prepared food to be overserved; we have supersized our home cooked meals too. An interesting study[19] out of Denmark tracked multiple editions of the same cookbook over 100 years and found that the average portion size in calories of a homemade meal has increased by 77%.

The problem is not so much the size of the burger or the number of fries; it's that we're eating it all — every last morsel. For a wide variety of reasons, psychological, physiological, sociological, environmental, and more, when presented with a ginormous number of attractively packaged calories, we will eat them. In Chapter 6 we will learn about some of the hormones that help regulate appetite, but they don't stand a chance against an Asiago Ranch Chicken Club (670 calories). It seems the only regulator on how much we will eat is an active desire not to overeat. In other words, unless your weight is front of mind, you'll eat what's in front of you.

[18] https://www.digitalcenter.org/wp-content/uploads/2018/12/2018-Digital-Future-Report.pdf
[19] https://journals.sagepub.com/doi/abs/10.1177/1403494813498468?journalCode=sjpc

Beyond the surplus of highly palatable and empty calories, the sedentary lifestyles, and the tendency of most people to overindulge, there are a number of other factors affecting America's waistline, including race, education, poverty, food deserts, and social norms. Unfortunately, these are all far outside the scope of this book. Suffice it to say — as is too often the case in America — unless you are reasonably affluent and unreasonably white, you have a much higher bar to clear when trying to get and stay healthy.

Spoiler Alert: It's the Sugar

Let's get back to our macronutrients and examine their individual impact on our weight and overall health, and while we're at it, dispel some myths and outdated information.

First up, protein. Are we eating too much protein? Can protein make you fat? This is the easy one. No, protein is not making you fat. While it's mathematically possible to become fat from eating too much protein, it's next to impossible to do so in actuality. A high protein diet may involve weight gain, but it will likely be from added muscle mass, not fat. Even if you're chowing down on protein bars and supplements, you're not likely to consume more protein than your body can effectively use. This is not the same thing as saying that it's okay to have hamburgers for every meal (though it is, if you make them at home and skip the bun), I'm just saying that in a well-balanced diet, you're likely not eating so much protein that you get fat from it.

Fat makes you fat, though, right? I mean, it's right there in the name. That shit'll kill ya. Eat too much saturated fat (meat and dairy) and it's lights out for you, sucker. Too much cholesterol (fat's little brother) and you'll stroke out right in the middle of 18 holes. Well, the "bad" cholesterol, anyway. Which one was that again? HDL? LDL? Whichever, doesn't matter, no more eggs.

Well, yes and no. When you are consuming too many calories, any excess fat will indeed make its way to your fat cells for storage (see Chapter 6 for a more complete description). It's also possible for a great deal of the "too many calories" to come from fat, mainly in the form of industrial fats added to packaged and prepared food. But if your diet consists primarily of whole foods, that is, *real* food, then it is more likely the case that you have too little fat in your diet, rather than too much, especially if you made the mistake of going on a low-fat diet.

I recently had reason to make a dish from Julia Child's *Mastering the Art of French Cooking*, which is basically ketogenic porn; every dish describe therein is saturated in saturated fat. As I mixed the melted butter and cream into the chicken stock, I got to thinking about the French Paradox. This is the term used for the seemingly inexplicable fact that the French population has a much lower incidence of coronary heart disease than US and UK populations, even though they have a diet that is high in saturated fats. Over the years there have been many attempts to reconcile this. Maybe it's because the French smoke more and that suppresses their appetites. Maybe it's because, after losing the Franco-Prussian war in 1870, the French instituted an early-life nutrition plan so they can field a more fit army. Or maybe, just maybe, it's the wine. Oh, we so wanted it to be the wine. Could be the polyphenols, you know, or the resveratrol, or something — anything? If there was just some kind of magic in wine that gave us a +5 saving throw against heart disease, then we could all eat like pigs, get drunk, and pay no higher price than a mild hangover. Alas, it wasn't the wine, nor was it the cigarettes or the Prussians. The answer to the French Paradox is much simpler and more evident than that: fat, it seems, does not cause heart disease[20].

Despite what we've had shouted into our ears for the past 50 years, an increasing body of evidence suggests that the consumption of fat — saturated or unsaturated — and cholesterol is not a significant contributor to heart disease, stroke, or metabolic syndrome (that last, by the way, is the name for a cluster of at least three markers for a future diagnosis of cardiovascular disease or type 2 diabetes. These markers include obesity, high blood pressure, high blood sugar or triglyceride levels, and low levels of HDL — that's the "good" cholesterol, so you know. An astonishing 25% of American adults suffer from metabolic syndrome). It turns out that natural fats are not killing us. In fact, eating as little fat as we do is a big part of what is making us increasingly unhealthy. Not just because we've replaced fat with sugar in our diet (which we have, and that's awful), but also because we are simply not getting enough fat to meet our metabolic needs. Indeed, a recent nine-year study[21] of nearly 500,000 Chinese adults found a dramatic decrease in cardiovascular disease in people who ate at least one egg a day. Another recent Canadian study[22] not only found that increased carbohydrate consumption, especially of refined sugar, led to a 28% higher risk of death, but that increased fat consumption led to a 23% decrease in mortality rates.

---

[20] There are likely other factors at play here too. The French, as a culture, tend to eat healthier foods and not eat as much processed or fast food, they consume smaller portions over longer periods of time, and they tend to dine socially.

[21] https://heart.bmj.com/content/early/2018/04/17/heartjnl-2017-312651

[22] https://www.thelancet.com/journals/lancet/article/PIIS0140-6736(17)32252-3/fulltext

I am not cherry-picking studies, either. Since the 1990s, study[23] after study[24] has found no significant link between dietary fat intake and coronary heart disease (CHD). Increasingly, they have found the opposite. Furthermore, and to the surprise of many, they are instead finding a causative link between carbohydrates and CHD. Not all carbohydrates, mind you; CHD is primarily linked to an excess of sugar and refined grains.

To be perfectly clear, I'm referring to natural fats and other lipids, the ones found in meat, dairy, nuts, etc. I am most certainly not referring to manufactured trans fats. That shit really will kill you, although no food processing company in the U.S. is adding trans fats to their products anymore, because the FDA mandated that they be removed by June 2018. Of course, the food companies have to replace trans fats with something, and Dow and Monsanto are stepping up to the plate with their engineered replacements. While these new fats do have a better health profile than the trans fats they're replacing, I'll be damned if I eat them. I am also not referring to the added non-trans fats found in processed foods, which are also industrially processed using various solvents, high temperatures, and deodorizers[25]. When I say that fat is good for you, I am only referring to the fats you consume as part of a healthy diet, even if that diet is biased toward high-fat foods.

This is not a license to start slurping down quarts of lard. While you should feel comfortable consuming natural fats, as always, keep things balanced. For your body to function optimally, you also need to eat grains and fruits and vegetables and fish and more. For the time being, it's also likely a good idea to favor omega-3s (fish, soy) over saturated fats (meat, dairy). When you are consuming beef, pork, chicken, and lamb, know that pasture-raised animals produce meat, milk, and eggs that are far healthier and tastier than their industrially raised counterparts. Finally, while evidence does seem to be trending toward the exoneration of fats as a contributor to cardiovascular disease, the matter is by no means settled and likely won't be for decades. You would be wise to follow your physician's advice.

Setting aside the issue of cardiovascular disease, etcetera, the question is: Does fat make you fat? It can, but it most likely will not. Not all by itself, anyway. Like protein, dietary fat is an important macronutrient. Unfortunately for fat's reputation, it shares a name with both body fat and the adjective describing overweight people. Even more unfortunately, it has been incorrectly demonized

---

[23] https://www.ncbi.nlm.nih.gov/pmc/articles/PMC2950931/

[24] https://www.ncbi.nlm.nih.gov/pmc/articles/PMC4532752/

[25] The health impact of the various vegetable oils found on supermarket shelves is highly suspect. As a rule of thumb, ignore all of them except for cold-pressed olive oil. If desired, cold pressed coconut oil is okay. Everything else, even the organic, heart-friendly, free range, artisanal oils, should be avoided.

by the USDA and the medical community for decades, even as the true cause of our corpulence was praised. When the USDA introduced the food pyramid graphic in 1992, the foundation was made up of the "bread, cereal, rice and pasta" group, suggesting 6-11 servings daily, and the top consisted of "fats, oils, and sweets," and an admonition to eat sparingly.

But fat, despite its name, doesn't make you fat. If you are consuming a healthy, balanced diet, then the fat you eat will be put to use as designed, either burned for energy immediately or stored for later use and metabolic regulation. The trick is simply to balance caloric intake with energy expenditure. Consume more calories than you burn, and you will gain weight. Full stop.

Now here is the part where I test your sense of humor: not only is fat not making you fat, if you were to consume more of it — a lot more — you will actually lose weight. We'll return to this seemingly outlandish fact several times in this book.

So don't knock fat and don't be afraid of cholesterol. The fat on your steak and the cholesterol in your eggs are not making you fat and they are not making you sick. Besides, fat is flavor. Lack of fat is the reason Julia Child has us reaching for the butter, oil, or cream when making chicken breasts. I, for one, don't want to live in the bleak and tasteless dystopia that a world without fat would surely bring about.

If it's not the protein and it's not the fat, then it must be the carbs making us fat, right? Again, yes and no. Like dietary fat, carbohydrates in and of themselves are not making us overweight. They are as important a component of our diet as their macronutrient brethren. Once more, it is an excess of calories that is making us fat. However, that excess of calories is coming predominantly from a single source, and that source is a carbohydrate. I'm talking, of course, about sugar.

Yes, it's the sugar and, to an equal extent, refined grains, that are almost singularly responsible for the world's obesity epidemic. Refined grains, by the way, are grains that have been milled so as to remove the nutrient rich bran and germ as well as the fiber, leaving just the starch (in essence, sugar) behind. These refined carbohydrates (typically with still more sugar added) are then made into donuts, white bread, and pasta — breakfast, lunch, and dinner, with nary a nutrient in sight. Admittedly, fat had a guilty look on its face, and the trans fats are every bit as bad as we thought, but it's sugar that's making us fat. Sugar is making us obese. Sugar is killing us.

That phrasing, however, is misleading; it ascribes to sugar a motive force that it most certainly does not possess. Saying sugar is killing us is like saying bullets are killing us, accurate as far as it goes, but at least four "whys" removed from the root cause. In the case of sugar, what I'm referring to is the added sugar (and starch) found in processed food. I am not referring to the saccharides found in cabbage, quinoa, or milk. No one ever got fat from eating too much cabbage. Nor am I referring to the more obvious sugars and starches found in foods like bananas, potatoes, and whole grain rice. I am referring to the simple sugars and starch served up to you every day in the form of soft drinks, french fries, and pasta.

Unlike (natural) fat and protein, it is trivially easy these days to consume too much glucose, and too much glucose, or too much anything that ends in "ose," will make you fat. That said, we don't have as much of an opportunity to OD on glucose as we once did, simply because in processed foods it has largely been replaced with its cheaper, sweeter, and government subsidized cousin, fructose, specifically in the form of high-fructose corn syrup (HFCS).

It's worth pausing here for a moment. You are likely aware that there is a chorus of voices — very well credentialed voices — that, when discussing the obesity epidemic, point the finger directly at fructose[26]. In particular, HFCS. You may be surprised to learn, however, that a number of equally well credentialed voices claim correlation does not equal causation and there's no good evidence linking fructose to obesity[27]. I, of course, have no credentials whatsoever, so it's not particularly relevant where I stand on this issue. If asked, however, I'm leaning with the correlation != causation crowd. Fructose is likely not quite the source of all evil that current headlines make it out to be. Your body can process fructose, if only indirectly. The real problem with HFCS is that there's so damn much of it.

It's too little movement combined with too many calories, primarily in the form of simple carbohydrates, primarily in the form of sugar — any sugar, glucose, fructose, or sucrose — that makes you fat. It's also increasingly clear that sugar is almost solely correlated with *all* of the ailments associated with the western diet: cardiovascular disease, diabetes, hypertension, and so on. Indeed, it's safe to say that all of these diseases are actually just symptoms of a single underlying disease, sugar-induced obesity.

Since diabetes, the scourge of the Western world, is so strongly correlated with obesity, it makes a useful proxy for obesity as a whole. As you likely know, there are two primary types of diabetes: type 1 and type 2[28]. Type 1 diabetes means that your pancreas is malfunctioning and is unable to produce insulin. It is an autoimmune disorder, and is not the result of a bad diet or lack of exercise. You

[26] https://www.ncbi.nlm.nih.gov/pmc/articles/PMC5409744/
[27] https://www.ncbi.nlm.nih.gov/pmc/articles/PMC4078442/
[28] We won't be discussing gestational diabetes.

cannot cure or roll back type 1 diabetes. You can, however, manage it. Not just by monitoring your glucose levels and taking your insulin shots, but by switching to a low-carb diet. More on this in Chapter 6.

Type 2 diabetes, however, that's a completely different beast. With type 2 diabetes — which used to be called adult-onset diabetes, but had to be renamed due to rising obesity rates in children — the pancreas is functioning normally, releasing insulin in response to blood sugar levels, but the cells are not listening to the insulin signal (we'll cover the science behind insulin and blood glucose in Chapter 6, as well). They have become resistant to insulin, and thus do not express the protein needed for glucose to cross the cell membrane and provide fuel for the cell. The scientific community is not 100% certain what causes type 2 diabetes, but there is no debate that it correlates strongly with obesity and a sedentary lifestyle[29]. Roughly 95% of all people with diabetes have type 2 diabetes, and 90% of them are overweight.

Here are the obligatory harrowing statistics: In 2015, nearly 10% of the U.S. population, some 30 million people, had diabetes (types 1 and 2 together, which is mostly type 2). Worldwide, that number balloons to well over 400 million. There are 1.4 million new cases of diabetes diagnosed in the U.S. each year, although that number has actually been going down since 2008. An estimated 34% of U.S. adults — fully one-third of the country, that's 84,000,000 people — have prediabetes, many without even being aware of it. Diabetes was the seventh leading cause of death in the U.S. in 2015, killing nearly 80,000 people. It is also the seventh leading cause of death worldwide, fourth if limited only to developing nations, killing approximately 3.7 million people each year, directly or indirectly. In the U.S., the diabetes epidemic is costing us nearly $330 billion per year, more than twice the annual cost of smoking- related illnesses. Globally, the direct healthcare costs of diabetes are a staggering $827 billion a year, not including the indirect costs related to lost productivity and reallocation of financial and healthcare resources, and almost all of these numbers are on the rise. Between 1980 and 2014, the number of adults globally with diabetes increased from 108 million to 422. In the U.S. over roughly the same time period, the number of adults diagnosed with diabetes also quadrupled, from 5.5 million to 23 million. Furthermore, it's estimated that by 2030, the losses to worldwide GDP due to the direct and indirect costs of the diabetes pandemic will reach a face-melting $1.7 trillion![30] And the sad thing is, nearly every diabetes-related amputation and every

[29] There are, of course, other predictors, including genetics, age, smoking, race, and whether someone is already managing a lifestyle-related disease, e.g. metabolic syndrome or fatty liver disease.

[30] U.S. health statistics: https://www.cdc.gov/diabetes/pdfs/library/diabetesreportcard2017-508.pdf
U.S. financial statistics: https://www.ncbi.nlm.nih.gov/pubmed/29567642
Global health and financial statistics: http://www.who.int/diabetes/global-report/en/

death and every dollar spent has been unnecessary. A sensible diet and sufficient exercise and — dare I say it — regulation and taxation of the processed food industry will save the planet untold levels of discomfort, death, and expense.

But don't wait too long for your country or state to start taxing sugar. Even California couldn't pull that off. They tried, but failed. Not that they lacked the desire or the political will, far from it, four California cities had already taxed sugary drinks. What happened was that Coca-Cola, Pepsi, and other sugar dealers spent $7 million getting an initiative on the ballot that would have made it nearly impossible for any municipality to raise taxes. Then they went to the state legislators with a proposition: you pass legislation making it illegal to tax sugar, and we will drop the ballot initiative. The technical term for this tactic is "extortion," but it's more commonly known as a shakedown. Having little choice, California's state senators acquiesced and passed the legislation that the soft drink lobby paid for. And if that doesn't just piss you right the fuck off, I don't know what will.

Chapter 3: Cut

My croissant was already gone. Denise's muffin sat uneaten and unloved on the plate before her. We dawdled for a bit, then went home. I thought nothing more about weight loss until after Labor Day. Early one weekday morning, at around step number 1,233, I realized that if I was going to exercise five or six days a week, this would be a good time to take some weight off. When I got home, I weighed myself: 199 pounds. I went downstairs and scrambled an egg. A little bit later, on my way to work, I came up on the Dunkin' Donuts and kept on driving.

I didn't Google "how to diet." I didn't follow a plan, I didn't meticulously count calories (though I learned what they are and how to estimate them), and I didn't join Weight Watchers or start buying diet foods. I did set a target weight, though, of 180 pounds. I thought that was the weight I was "supposed to be." The weight where all the visible excess fat would be gone, but I wouldn't be scrawny. I was wrong about that, as we'll see, necessitating a second diet a few months later. For now, let's discuss the first 20 pounds.

Despite not knowing what I was doing, getting to a healthy weight is not rocket science, and I was as aware as everyone else that eliminating added sugar was a good place to start. I applied some commonsense principles and some willpower, and lost some weight. If you are hoping to drop a few pounds, I would be surprised if you don't get the same results. Here's how I did it. There are only three rules.

Rule Number 1: No More Processed Foods

Rule number one, the most important rule: eliminate processed foods from your life. If you do nothing else, do this. You can save the exercise and caloric restriction for another day, but you need to change your diet *now*. Doing so will, at a stroke, cut your sugar, salt, fat, and caloric intake dramatically. You will start feeling better, and the pounds will start to slip away. You've heard this admonition a thousand times before, but maybe it wasn't the right time. Maybe now is not the right time either, but when you're ready, you'll need to drop the bulk of processed foods from your diet.

Rule number one does not refer to weight loss *per se*, it refers to your *diet*, the foods you eat all the time, regardless of whether you're actively trying to lose weight or not. You need to drop the so-called "Western diet" of processed foods that are high in sugar, fat, salt, and not-immediately-lethal chemicals, and adopt a healthy diet of real food: fruits, vegetables, and whole grains, as well as fresh meat, fish, and dairy. Even if you never intend to lose a pound, you need to start eating better. However, if you do cut out junk food and replace it with natural food, you will almost certainly find that you'll lose weight anyway, simply because of the difference in calories between, say, two eggs (roughly 150) and two donuts (roughly 600).

Start with the sugary drinks. Soda, fruit juice, sports drinks, and so on, they're all out. Gone. Never to be seen again. Coke may be "the real thing," but a 12 oz can contains 140 calories and 39 g of sugar[31] that will hit your liver like a freight train. Gatorade may be "made for athletes," but it's not made for the rest of us. A typical 32 oz bottle contains 56 grams of sugar. Better than soda (when measured by the ounce), but unless you've just finished a triathlon, you shouldn't drink it. Snapple may be "made from the best stuff on earth," but a 16 oz bottle has 160 calories packed into its 39 g of sugar[32]. Cranberries have a lot of nutritional value, and you should definitely eat them, just don't drink them. A single cup of cranberry juice has 28 g of sugar. And please don't get me started on Frappuccinos (13.7 ozs / 45 g), Vitaminwater (20 ozs / 32 g), and Red Bull (12 ozs / 40 g).

You may be wondering if diet soft drinks would make a good substitute. To which I would reply, I suppose so, if your only goal is weight loss. My question back to you is, do you really need something cloyingly sweet to drink all the time? Me? I'd choose a 300-calorie pint of milk over a zero-calorie Diet Coke any day. But honestly, I'd really just have a glass of water.

Which brings us to what you should be drinking instead: milk and water. That's pretty much it. So you know, water doesn't have to be boring and

---

[31] That's what it says on the can, anyway. If you do the math, 39 grams of sugar times 4 calories per gram should supply 156 calories of energy. The Coca-Cola Company says the discrepancy is due to their use of the more exact measurement of 3.68 calories per gram of sugar (rounded down). Pepsi and Dr. Pepper are guilty of the same sleight of hand.

[32] At least Snapple isn't playing fast and loose with the number of calories in a gram of sugar, even though it's made by the same company that makes Dr Pepper.

flavorless. I get the bulk of my water in the form of (black) coffee. If you like club soda, even the lightly flavored kind, that's fine. Mineral water is especially good. Unsweetened tea, iced or not, is also good. Drinking water doesn't have to come straight from the tap, although here on the East Coast, at least, tap water is generally safe, tasty, and cheap.

As far as milk goes, yes, it is high in calories and fat, and it contains sugar in the form of lactose. But, unless you're lactose intolerant, your body knows how to process that sugar, breaking lactose down into glucose and a previously unmentioned monosaccharide, galactose, so that's not an issue. Plus all that fat? It's good for you, remember. There's a ton of protein in milk too, not to mention an abundance of minerals, such as calcium and phosphorus, and vitamins D & A. Indeed, milk is a veritable superfood. Also, and this is important, milk is delicious. Whole milk, that is, not the tasteless gray abomination that is skim milk. Besides, what else are you going to have with a peanut butter and jelly sandwich?

What about alcohol? Ideally, you should abstain from drinking altogether. As mentioned earlier, alcohol has seven calories per gram, three more than sugar. Wine, beer, and certain cocktails have additional calories from the carbohydrates they contain (distilled spirits have no carbs). However, complete abstinence from alcohol is a bit much to ask. Instead, if you are a light drinker, consuming, say, the equivalent of one glass of wine a day or less, then feel free to continue imbibing. Besides, there's ample reason to believe that a modest amount of alcohol is actually good for you[33]. As you know, I drank considerably more than one drink a day while I was dieting, but I clearly shouldn't have. Do as I say, not as I do, and all that.

Sugary drinks are not the only offender, they're just the worst. Refined grains are next. Recall that refined grains, such as white flour and white rice, are grains with the fiber and nutrient rich bran and germ removed, leaving mostly starch behind. Further recall that starch is just a lot of glucose stitched together. From your body's point of view, starch is nearly indistinguishable from plain sugar. Cutting out refined grains means that cookies, muffins, donuts, and bagels are gone. As are croissants, pies, cakes, white bread, pizza, pancakes, and most breakfast cereals. I can't and don't need to enumerate every possible baked good; you know what they are. Don't eat them. Eliminating refined grains also means eliminating most pastas as well as white rice. While you're trying to lose weight, it's best to reduce carbs across the board, but to be clear, whole grains are still an

[33] There's also ample evidence suggesting that no amount of alcohol is good for you. But I put that here in a footnote on the chance you'll overlook it. You can thank me later.

important part of your diet. You can find whole grain bread, pasta, rice, and other products in the same aisle as you find their refined counterparts. Eat them instead.

Confections are obviously on the no-fly list. Consequently, no more candy or ice cream. Chips and other snack foods are also a no go, so forget the potato chips, pretzels, Cheez Doodles, and Fritos. I think it's obvious that fast food is out. In fact, you want to limit prepared food as much as possible. As a rule of thumb, the less expensive the meal, the more it is to be avoided. Finally, there's all those premade meals, side dishes, and condiments: Hot Pockets and frozen pizzas, tater tots and Rice-A-Roni, salad dressing and mayonnaise. All that has to go too.

Those last few paragraphs had a lot of imperatives: don't eat this, don't drink that, don't ever enjoy yourself — forever! But like temperance, that's a pretty tall order, and not actually necessary. A world without pizza and gelato is about as attractive as world without fat. It is not required to completely eliminate all of these yummy foods from your life, you only need to remember that they are treats, not staples. While you are actively slimming down, it's best to dial back on all of these sugary, starchy foods — ideally to zero. But once you're at a healthy weight and sufficiently active, then by all means, have an occasional slice of pizza.

Similarly, others will tell you that starchy foods — potatoes, beans, rice and other grains — are to be avoided, but I disagree. These are real foods with lots of nutritional value. You don't need potatoes every night, especially in french fry form, but a baked potato once a week is not a problem. If you have white rice as often as you have spinach, you'll be fine. Better yet, switch to whole grain rice and have as much as you care to. For that matter, lentils, barley, quinoa, and bulgur wheat are all delicious and rich in fiber and nutrients to boot, so eat up.

But enemy number one is sugar. Refined sugar. Added sugar. As with starchy foods, do not stress even a tiny little bit about the natural sugar found in an apple or an orange. If you want an apple, eat an apple. There are about 100 calories in an apple, nothing to sneeze at, but there's also a lot of fiber to help you feel full and aid your overall digestion. An apple also has a decent amount of vitamin C as well as antioxidants. With its low glycemic index, an apple will dribble sugar into your bloodstream at a rate your body can easily process, delivering energy for a longer period of time and avoiding insulin spikes.

In contrast, something like Froot Loops, which is basically sugar and food coloring held together with a bit of (refined) corn flour — will charge into your bloodstream like a bat out of hell. There is no fruit, or indeed, froot, to be found anywhere near the things. Even though Kellogg's adds a few vitamins to Froot Loops, they're a "food" that has pretty much no nutritional value at all. Despite that, this is what it says on Kellogg's website:

> *Kellogg's Froot Loops™ is packed with delicious fruity taste, fruity aroma, and bright colors. Made with whole grains and lightly sweetened, Froot Loops is a fun part of a complete breakfast, and is a good source of fiber.*

Lightly sweetened! It contains more sugar than any other ingredient. Speaking of sugar, here's what Kellogg's has to say about that:

> *Sugar in cereals — including kids cereals — contribute [sic] less than 5 percent of daily sugar intake. Yet it adds taste, texture and enjoyment to cereal, while encouraging the consumption of fiber, vitamins and minerals — essential nutrients that you and your kids might not otherwise get from any other meal.*

That's "5 percent of daily sugar intake," mind you, not 5 percent of the *recommended* daily intake. (I downloaded the paper[34] they got that stat from, by the way, and I can't find any such claim. But other sources confirm that, yes, breakfast cereal is not where people are getting their excess calories from; they're getting most of them from soda. That's still not a good thing for Kellogg's, however). The World Health Organization recommends that adults consume no more than 6 teaspoons of sugar a day. A single bowl of Froot Loops has 10.5 teaspoons in it. If you want your children to get "essential nutrients" with taste, texture, and enjoyment, then give them scrambled eggs, an apple, and a glass of milk for breakfast. They'll like it more, and it's good for them.

I don't mean to pick on just Kellogg's. Pretty much the entire processed food industry is at fault for loading you up with sugars, salt, trans fats, and preservatives, then lying to your face about the consequences. Kraft and Nestle and General Mills and Tyson and so on are just as guilty of preying upon children, the uninformed, and harried parents. It's infuriating.

But I digress. Everyone knows that Big Food doesn't give a damn about your health. The processed food industry of today is like the tobacco industry of the seventies, promoting deeply unhealthy products and telling you they're good for you, or at least not that bad. But they are not good for you, they are that bad. Sugar is the new tobacco. If you want to stay away from sugar, then stay away from processed foods.

That said, in today's society, you can't reasonably stay away from all processed foods. In fact, I don't think we should. I can make a peanut butter and jelly sandwich at home, but I'm not about to grind the peanuts, preserve the grapes, or bake the bread — and I don't even have a cow to milk. Fortunately, there are ways to go grocery shopping that will help you limit your consumption of

[34] https://www.ars.usda.gov/ARSUserFiles/80400530/pdf/0102/usualintaketables2001-02.pdf

processed food to the healthiest essentials. First, if at all possible, do not buy food in supermarkets (not to mention gas stations or big box stores). Instead, look for a local grocer or butcher. If supermarkets are the only option — and sadly, that's probably the case — then use the Michael Pollan[35] algorithm of shopping around the perimeter. That is, prefer the produce, meats, and dairy and skip everything else. But don't lose any sleep over buying frozen or canned fruits and vegetables. Never in my life have I seen a lima bean in its native form, and I eat lima beans all the time.

Of course, there are still some things that you can't (milk) or won't (sandwich bread) make yourself. In that case, go for the rule of five. Look for products that have five or fewer ingredients; the closer to one, the better. That means bacon is in, but Kraft Stove Top Stuffing is out (28 ingredients, the second most common being high-fructose corn syrup).

Some other good rules, not necessarily related to weight loss: No products with ingredients that you can't easily pronounce (e.g. disodium inosinate — also Stove Top), no products with ingredients you wouldn't use in your kitchen (e.g. butylated hydroxytoluene, BHT — Kellogg's Corn Flakes, which works for the previous category, as well), and no foods that would normally spoil without refrigeration and they're not refrigerated (e.g. mayonnaise).

In short, eat real food. Or, as Michael Pollan, would have it: "Eat food. Not too much. Mostly plants." Though I admit to dialing up the meat and dairy.

Rule Number 2: Reduce Portion Size

The good news is, if you reduce the amount of processed and prepared food in your life, you pretty much don't have to worry about the meals you (yes, you) are preparing instead: they *will* be healthy. If you don't have a frozen pizza for dinner, but instead have a nice fat-laden porkchop, with brussels sprouts drowning in olive oil, and some quinoa with mushrooms sautéed in butter, you'll have consumed roughly half as many calories as are in the pizza and no refined carbohydrates. You'll also get all the healthy fats, plus the natural vitamins, minerals, proteins, and other goodly things. It will take 45 minutes to prepare that meal, most of which is waiting on the meat to cook through. Make two pork chops, and you have lunch for the next day.

That pork chop dinner may sound like a lot of food, and it can easily exceed 1,000 calories, but it doesn't have to. A more reasonable portion size weighs in at

[35] https://michaelpollan.com/

around 750 or even 500 delicious, healthy calories. Whereas just three slices of DiGiorno's Original Four Cheese Pizza top the scales at 870 calories. Despite the fact that not all calories are created equal, weight loss is still a numbers game. If you consume fewer calories than you use, you *will* lose weight. This is something you already know, but let's run the numbers anyway.

A six-foot one, 53-year-old, 200-pound sedentary male (that is, me in April 2017), burns about 2,500 calories a day just being alive. Since a pound of body weight equals about 3,500 calories, for this not-so-hypothetical man to lose one pound every week, he needs to reduce his caloric intake by about 500 calories a day. Consume even fewer calories and lose weight faster. For me, those 500 calories were bound up in my daily intake of cookies and donuts.

Let's say your diet doesn't have any obvious empty calories in it, then what? Now you just need to eat less. Take what you would normally eat in a healthy breakfast, and have about a third of it. Have half of what you would normally eat for lunch, and three quarters of what you would otherwise have for dinner. Or do it the other way around, doesn't matter. Eat real food, and, until you hit your target weight, eat less of it.

On the first day of my weight loss effort, as I've already mentioned, I had an egg for breakfast. That's it. One egg. Since I had no plans to start losing weight until I started losing weight, I had no lunch to bring to work. I was forced to go to the local greasy spoon and get a sandwich. I got a chicken salad wrap and threw half of it away. I also skipped the chips and juice, and had a bottle of water. I don't remember what I had for dinner, but I do know that I had roughly ¾ of what I normally would have had.

Some educated guesses and back-of-the-envelope calculations put my calorie intake on day one at about 1,200 calories (I assumed a good dinner), a full 1,300 less than maintenance level. I repeated that the next day, and the next. I switched things up in the days that followed. Some mornings I'd have two eggs, maybe with a single slice of whole grain toast (yes, with butter — and even jam). Some mornings I'd have half a grapefruit and some cottage cheese with blueberries. Breakfast was easy, filling, and tasty, and probably never exceeded 250 calories.

After that first day, barring extenuating circumstances, lunch was always leftovers. There's a number of benefits to this: you save a ton of money, like $50 a week; it's always "real" food; and you control portion size. Let's say I had salmon, asparagus, and (whole grain) rice the night before. For lunch, then, I'd have maybe four ounces of fish (roughly the size of a deck of cards), four or five spears of asparagus, and a few ounces of rice. That's a good lunch and only about 400 calories, but let's be generous and say lunch was 500 calories a day on average.

Dinner might be an 8 oz steak, corn on the cob, and spinach, all of it swimming in butter. That's about 1,000 calories, which is a pretty substantial dinner, and probably above average for what I was consuming. But let's run with it. That means that on average, while dieting, I consumed 1,700 calories per day, 300 calories fewer than target and 800 calories fewer than when I was not dieting. In actuality, it's likely that my daily caloric intake was closer to 1,500 (not including the wine).

Did you see what just happened? I was eating less without starving myself, even as I enjoyed three delicious meals a day. Just switching away from junk food and dialing back the portion size was enough to bring the calories under control, while providing my body with all of the nutrients it was missing. Which, among other things, made my body a better calorie processor. It's a virtuous cycle.

Rule Number 3: Exercise

Diabetes, obesity, high blood pressure, and heart disease are very much Western diseases, (although that's changing as the processed food industry, like the tobacco industry before it, starts aggressively marketing its products in the developing world). Besides the vats full of fat, sugar, and salt found in processed foods, the other driver behind these diseases is a sedentary lifestyle. In addition to switching your diet, if you were to walk for 30 minutes a day, Western diseases wouldn't be a concern of yours. Frankly, if you eat well and exercise regularly, you won't need to worry about your health much at all.

Which takes us to our final rule, rule number three: You must exercise.

First, know that you will not lose weight by exercising. Or rather, you won't lose weight *just* by exercising. You won't be able to burn off enough calories with exercise alone to compensate for not eating well or eating too much. Remember that 100-calorie apple? You might randomly pick up that apple and eat it, which is totally fine. But in order to burn 100 calories, you have to walk or jog a mile[36]. A whole freaking mile. For one apple! There's really no way to keep up. You cannot exercise away the calories you consume. If you want to lose weight, you must reduce your caloric intake.

Why then is exercising a rule? Two reasons: One, even if you can't exercise enough to compensate for an excess of calories, it still counts. When you're actively losing weight, you need to be in a calorie deficit, that is, burning more

[36] That's an average number. You can be more precise with calculators such as this one: https://www.runnersworld.com/training/a20801301/calories-burned-running-calculator/

than you consume. In the case of the middle-aged, slightly pudgy, rather torpid man discussed earlier, he needed to consume 500 fewer calories a day in order to lose (roughly) a pound a week. There are three ways to go about that. He could stop eating every morning, he could jog five miles every morning, or he could have a decent breakfast and go for a three-mile walk.

He'd be successful in his diet if he just skipped the donuts, but without replacing them with something else, he wouldn't have any energy in the morning. He could do the jogging thing, but five miles a day is quite the commitment and likely not even possible. Or he could replace a 500-calorie donut with two 75-calorie eggs and a negative 300-calorie walk. That is, fasting reduced his daily calories by 500, a five-mile run also reduced his calories by 500, but changing out the donut for a nice breakfast and a pleasant walk reduced his caloric intake by 650.

The other reason is that exercise, even a modest amount, is enormously beneficial for body and soul. You may think, as I used to, that exercise is going to make you feel anything but positive. You might not believe what I had to say about boot camp being rather agreeable. Fine. But you need to know that exercise isn't a chore. I don't know where I ever got that notion. Maybe grade school PE class. Maybe because I'm inherently lazy. Exercise does not mean climbing ropes and doing squat thrusts (unless you want it to), it just means getting your heart rate elevated for 30 minutes a day. Do you have a dog? Take it for a walk two or three times a day. Or take that three-mile walk every morning. Go play tennis or ride a bike or build your own cardio workout. It doesn't really matter. I like my cardio outdoors, but if you don't (or can't), then treadmills and ellipticals are just as good.

The positive impact of exercise cannot be overstated, and I'll have a lot more to say about that in later chapters. Provide your body with the right nutrients and some physical activity, and it will start performing as designed, firing on all six cylinders. Exercise will reduce the risk of stroke, heart disease, diabetes, hypertension, and dementia. Exercise will reduce — even roll back — the effects of aging. Your skin will look better, you'll be more limber, even your vision and cognitive processing will improve.

And you'll feel it. You will simply *feel healthy*. And feeling healthy will make you feel happy, which will make you want to exercise some more. I'll return to this in the next chapter, but soon you will begin to define yourself as athletic, and athletic people watch what they eat. It's this change in mood and perception that makes exercise rule number three. Everything's better when you're happy. Exercise adds this positive feedback loop (and helps some with the calorie thing).

To sum up:

- Eat real food.
- Reduce portion size.
- Increase your activity level.

In other words, the not-so-secret to weight loss is diet and exercise. The good news is, it's not at all onerous. In fact, it can be quite pleasant.

The Diet that Works for You

Some readers may be making that face right now. The face that says, "Easy for you, maybe, but I've done all those things a hundred times and I can't keep the weight off." I get that. It *was* easy for me (at least those first 20 pounds were), but it's not easy for everyone. Just know that if you are overweight it is not due to laziness or some intrinsic moral failing. In fact, it's largely due to the typical Western lifestyle of cheap calories and long commutes. I watch people I love struggle with their weight every day. It's clear to me that dieting is not just effort and willpower. What I am saying, though, is that a successful weight loss plan will always contain the trinity of healthy food, caloric restriction, and exercise.

On a similar note, I am not here to fat-shame anyone. I was unhappy with my weight, I was unhealthy because of it, and it was easy enough for me to lose it once I decided to. But you may be perfectly happy with your weight, even if it's above what's currently considered appropriate. Few of us will ever look like a fashion model or a movie actor, and we shouldn't feel pressured to do so. Also, it is most definitely possible to be "fat and fit" (just as it's possible to be "slim and sick"). If you're eating well, getting regular exercise, and not already in the throes of metabolic syndrome, then by all means, enjoy a few extra calories and a few extra pounds.

That said, if you repeatedly struggle with weight gain, you may find yourself bouncing from popular diet to popular diet; Atkins this, Subway that, "Lose 25 pounds eating nothing but herring!" If so, I'd recommend that in the future you ignore every magazine article that promises a new way to lose weight, and every headline about the latest scientific discovery that will take off the pounds. And please, please ignore that one small trick you see on every website. Fad diets and crash dieting do not work.

To name names, the Paleo Diet has its heart in the right place, emphasizing foods that humans have evolved to eat and avoiding processed foods. Nonetheless, it is hard to recommend, for a variety of reasons. One, it dials back too far on fat, emphasizing lean meats (even though Paleolithic Inuit, for one, must have consumed a great deal of fat). Two, it's arbitrarily restrictive, allowing nuts, for instance, but not beans. Three, it fails to acknowledge that humans have continued to evolve in the last 12,000 years. For instance, many populations, notably in Western Europe, have evolved the ability to digest milk (lactose) beyond childhood[37], a feat that is likely due to the rise of post-Paleolithic pastoralism and dairy farming. Finally, the Paleo Diet disallows too many perfectly reasonable foods — including peanut butter! — that simply make the diet so onerous that it is not likely to be practiced faithfully or for long.

Similarly, the Whole30 diet starts off promising, recommending moderately sized portions of whole foods, but then it goes completely off the rails by suggesting that you purge an arbitrary list of foods from your diet for an arbitrary amount of time – 30 days. That list includes fiber and nutrient rich whole grains and legumes (what does everyone have against beans?!), as well as all dairy products. It's utterly ridiculous. I'm sure people get short term results from it, but like the Paleo diet, it's overly restrictive and completely irrational. That said, I bet they're making bank selling Whole30 cookbooks, meal plans, and coach certifications.

The Blood Type Diet is beyond stupid, so don't do that.

Intermittent fasting won't kill you, but there are easier and better ways to limit calories that don't make losing weight unpleasant.

You can be successful with the Atkins and South Beach diets. Any diet that dials back the refined carbs is heading in the right direction. However, I caution against spending any money on these, or trying to be religious about what to do in week one and what to do in week four. Dieting is a matter of consuming fewer calories, especially sugars, and exercising more. That's it. It shouldn't cost you anything and it shouldn't make you crazy. The hard sell from the South Beach Diet company to buy their premade meals is particularly worrisome. The health profile of these processed foods is modest at best, given that you can prepare fresher, tastier, more nutritious meals for the whole family for less money and with hardly any effort.

South Beach's parent company, Nutrisystem, does the same hard sell and is just as worrisome. And everything about Jenny Craig screams scam! I'm less concerned about Weight Watchers, but still can't see the need to pay for special foods. Shopping and cooking can be fun and easy, so long as your commute allows it (if it doesn't allow it, consider cooking batches on weekends and freezing them); you don't need to eat packaged frozen dinners, even diet frozen dinners. Remember, diet food is processed food, and none of these companies stand to make any money from you if you're successful in your diet.

---

[37] Which does not make Western Europeans superior in any way whatsoever.

One named diet that deserves special attention is the ketogenic (keto) diet. This is not a corporate sponsored thing, it's a diet in the sense that I defined it above: the foods you eat on a regular basis. Nor is it a fad; it's been around for almost a century, and is routinely used by the medical community to help control epileptic seizures in children. It is, however, only recently popular as a mainstream lifestyle choice that can also lead to weight loss. The keto diet is a low-carb diet — an *extremely* low-carb diet. It recommends that people consume only around 35 grams of carbs per day, 10 of which should be fiber. That's roughly the number of net carbs as are in a single potato or a pint of milk. Which raises the question: what are you to replace the carbs with? After all, you still need the calories. The answer: fat. Put another way, someone on a keto diet is expected to get around 70% of their calories from fat, roughly 20% from protein, and the rest from carbs.

If you're like everyone else to whom I've described the keto diet, you just did a spit take. Seventy percent of calories from fat is a strikingly high number! Or so it sounds to those of us who have been told for most of our lives to limit fat consumption. Even though the USDA ditched the food pyramid years ago, and its current recommendations[38] are a lot less simplistic than they use to be, they still emphasize fat-free or low-fat dairy and no more than 10% of daily calories from saturated fat. Given that the keto diet completely inverts this, you are justified in spraying milk (or water) across the pages of this book. Almost. What I find interesting is that no one's ever freaked out by the notion of getting 70% of their calories from carbohydrates. Because you should be. As mentioned above, the evidence continues to proliferate that it's the added simple sugars, and not fat, that is linked to America's epidemic of obesity, diabetes, and heart disease.

For the last 50+ years, the U.S. has conducted a nationwide, uncontrolled experiment on the impact of a diet rich in carbohydrates and low in fat, and the results are in, brother: we all got fat and sick. In contrast, the ketogenic diet is not an experiment. As noted, it's been around for 100 years, possibly as long as 2,000 years[39]. The ketogenic diet is proven and it is healthy[40], and it goes against everything we've been told for the last half century. Besides being effective in controlling brain seizures and other neurological issues, such as Parkinson's and multiple sclerosis, it is also effective in dialing back many of the symptoms of the Western diet. That is, it's therapeutic when it comes to obesity, heart disease, insulin resistance, diabetes, and so on[41].

---

[38] https://health.gov/dietaryguidelines/2015/resources/2015-2020_Dietary_Guidelines.pdf
[39] https://www.biblegateway.com/passage/?search=Matthew+17%3A14-21&version=NKJV
[40] https://www.ncbi.nlm.nih.gov/pmc/articles/PMC2716748/
[41] https://www.ncbi.nlm.nih.gov/pmc/articles/PMC3826507/

Relevant to you, however, the ketogenic diet is extremely effective when it comes to weight loss. This is likely because the high fat intake (now proving to be perfectly safe, remember, even desirable) and changes in the release of appetite-regulating hormones appears to mitigate against the hunger associated with other diets. Should you switch to a ketogenic diet, while cutting back the carbs and amping up the exercise, you can expect to drop about ten pounds a month. Easily.

You may wonder if I'm on it? No, I am not. Not currently, anyway. But that's not due to any perceived adverse health effects. It's only because I like my PBJs, and you can't have a PBJ without the J and the bread to spread it on, not to mention the milk to wash it down. I also occasionally have a bowl (or two) of Cheerios, and sometimes I have toast with my breakfast, and occasionally pasta and grapefruit and ice cream and all the things I listed above. All, I should note, in moderation, but not ketogenic moderation. I simply don't have the emotional fortitude to go full keto.

But I know someone who does: my wife. Denise has been in a permanent state of ketosis for eight months now. To hear her tell it, the keto diet has completely changed her life. She's lost 40 pounds, even though she barely thinks about calorie restrictions. More importantly, she's not hungry between meals, and consequently not anxious about food. She's full of energy and more fit than at any time in her adult life. She has not the slightest desire to ever move off this diet. She does not need pizza, french fries, or beer. She does need ice cream, of course, but she can whip up a batch of stevia-based ice cream in a flash. Similarly, she can turn almond flour into hamburger buns faster than I can turn ground beef into hamburgers, and she can turn cauliflower into just about anything. Most importantly, she had recently been diagnosed with prediabetes, and now that's completely reversed. Gone!

Despite Denise's success and, as you'll soon see, my own, I am not recommending a ketogenic diet for life. Even though the diet is scientifically sound, it is, like the Paleo and Whole30 diets, extremely onerous. Allow me to repeat, carbohydrates are not bad. For proof, take a look at any vegetarian; they eat nothing but carbs, yet tend to be both slender and healthy. It's the *excess* of *refined* carbohydrates found in processed and prepared food (soda, pasta, donuts, etc.), that we need to be wary of.

Here then is my recommendation for you. Select a target weight appropriate for your height and build. If you're not sure what that should be, then use the weight given towards the middle of the BMI chart linked above. For me, that's around 170 pounds. Because the ketogenic diet is safe and so remarkably effective at weight loss, I recommend that you get on it and stay on it until you reach your target weight — but not forever. Once you've reached your goal, gradually reintroduce minimally processed carbs back into your diet: milk, fruits, vegetables, grains, seeds, legumes, and so on. But not over-processed foods, the injunction against them still stands. Note, that switching the percentages of fat and carbohydrates in your diet does not change the overall calorie count, in order to lose weight, you must still reduce portion size, which will be easier on the keto diet. Similarly, you must still exercise regularly, not just while you're actively dieting, but forever. That's it. One paragraph. Practiced faithfully, you should be able to lose about ten pounds a month, without anxiety and without succumbing to the urge to cheat or quit.

Knife and Fork

So I snuck up on myself and cut way back on the calories. The effect was immediate and dramatic. On day two, I was down a pound. On day three, down another. By the fourth day I was at 196.

"One hundred and ninety-six pounds this morning," I said to the wife.

"That's just water weight, dear. It doesn't count." Turns out water weight is a thing. I did not know this.

At 196 pounds, everything just stopped. For five maddening days, I did not lose a pound. Then — click — 195. Click, click, 194, 193. Click, click, click, I hit 189 on June 19, just three weeks after starting.

"I hit 189 today," I crowed to the missus.

"Fuck you," she replied, demurely. Denise has been very encouraging during this whole transformation thing, but not so much about the weight loss.

But at 189, I plateaued again. For another week or so, I just bounced around either side of 189, until, eventually, the weight started trending down again. Then I hit 181, and 181 was rough. Just one pound away from my goal and I flatlined. For two whole weeks I could not break 181. On a couple of days my weight actually went up! For one week straight, my weight wouldn't even change fractionally, it was just 181.0 day after day. I thought the scale might be broken. It was quite distressing.

Distressing enough that I actually cut back dramatically (for me) on the wine for a few nights, and doubled down on the exercise. Finally, on August 2, the scale clicked over to 180. But that didn't count in my book; I need to be *below* 180 pounds for at least two consecutive days (to compensate for the damn water weight). On August 3, I was 179.4, and on August 4, after 67 days of caloric restriction, I was 178.6.

I've tried to make weight loss sound pleasant and fulfilling, and it can be, but it is not effortless. You need to drive back the desire to just cave, if only for one night. A variety of real food, the ketogenic diet, and exercise will go a long way toward helping you press on. Momentum helps too. "I've come this far, I'm not gonna quit now." But in the end, you must be mentally prepared.

I know this because I should not have stopped at 180. As I mentioned, 170 pounds (including some newly added muscle) would get me to a very healthy 10-13 percent body fat. But, hell, after two months of dieting, 180 was good enough. I upped my calories to sustenance levels and was able to keep my weight steady at 180 with little effort. But as the summer switched over to fall, I became increasingly unhappy with the love handles still showing on my waist. In late October, I decided to have one more go at dieting in order to get down to 170. My plan was simple; I was already eating healthy and exercising regularly, all I needed to do was cut back on the calories. I figured I'd diet until Thanksgiving, a month away, then celebrate with a feast.

Things went well at first, but I was never able to push my weight below 175. Even though I was able to easily drop 20 pounds a few months earlier, I was not able to repeat that feat now, and the only difference between the two attempts was my mental state. My heart simply wasn't in it, and it seems if you're not all in, you're not in it at all. Consequently, I cheated. I'd have my one egg for breakfast, sure, but maybe a bowl of (high fiber) cereal too, and, oops, probably packed more than I needed for lunch today. What's that, kids? Pasta again tonight? Well, okay.

Thanksgiving arrived and I ended the diet defeated. Not long after, winter arrived too, making it increasingly difficult to get enough aerobic exercise in. I decided to hunker down until spring, then take another shot at losing 10 pounds when the weather improved.

Having made this plan, I immediately broke it and started another weight loss effort on the day after Christmas. This was not smart. Three weeks later, finding myself all the way back at my starting weight of 180, I once again threw in the towel, never having broken the 175-pound barrier. The very next day, over coffee, I lamented about my latest failure to Denise.

"Oooh," she said. "Do you want to go on a ketogenic diet with me?"

"Go on a what?" I asked.

Denise explained the keto diet to me, occasioning me to retreat to the web, where I learned much of what I wrote above. In the end, just two days later, I tried once again to lose those ten pounds, but this time ketogenically and with Denise. While it remains true that you should never start dieting in winter, it's also true that if you're going to, you should do it with someone else. For me, at least, this was the extra push I needed to really, truly dial the calories back.

The results were dramatic. By January 26, just 12 days in, I had fought my way down to 173 pounds. I went through the keto flu (Google it) and bouts of insomnia (though I'm not sure if the sleeplessness was related). Then I stalled once again, not, I think, for dietary reasons, but because my body was trying to conserve energy. Soon enough, however, my weight started going down once more, and on February 10, 2018, less than a month after starting on the keto diet, I hit 170 pounds.

Immediately after reaching my goal, however, I reached for a pan and made myself a truly awe-inspiring plate of liver and onions, with bacon, barley, mushrooms, and dried cherries (Denise and the children were away). It was delicious. Despite having prepared that meal using an entire stick of butter, the presence of barley is your first clue that I was no longer ketogenic. Indeed, no longer dieting. Unfortunately, unlike the previous summer, I was not able to maintain my desired weight, and within a month I was a soul-crushing 180 again.

Spring arrived a month late in New England, but when it did I redoubled my aerobic exercise and launched attempt number four to get to 170 pounds and stay there. Once again, the initial pounds sloughed off easily, and once again, I plateaued for weeks, this time at 173 pounds. In mid-May I quit the diet in disgust, but then, without even trying, the last three pounds slipped away in as many days. It's late July as I write this, and I have been able to maintain this weight, plus or minus two pounds, without much effort for two months.

To my mind, the difference between failing to maintain 170 pounds in February and succeeding in May is simple: it's the weather. Completing a diet in winter seems to be just as difficult as starting one. It is dark and cold outside, and your primitive, mammalian, atavistic brain wants nothing more than for you to stay warm and full. For me, it was too difficult to maintain regular cardio exercise and too easy to choose pasta for dinner. But now it's summer again. The days are warm and the nights are short, and that means I am absolutely crushing it on cardio, burning through 3,200 calories a day over the last four weeks. Who knows what next winter will bring?[42]

---

[42] It takes a long time to write a book, so I can answer this now; it brought ten more pounds. But I'm not at all concerned, I'll run my way back to 170 come spring. Besides a few of those pounds are straight up muscle.

Chapter 4: Breathe

I don't remember my last cigarette anywhere near as well as I remember my first. This may seem odd given how momentous it should have been. It was, after all, the last of the roughly 270,000 cigarettes I've smoked in my lifetime. The last gasp, as it were, after 37 years of smoking. And yet, I have little recollection of that smoke, only that it took place on my front porch on the evening of July 11, 2017.

I have a simple fantasy of what that cigarette was like. In my imagination, as it was in real life, the weather is warm, and I am sitting on the porch. I pluck the last cigarette from the pack, run it under my nose like a fine cigar, and place it between my lips. In one fluid motion born of long practice, a lighter materializes. A flame is lit and touched to the business end of the cigarette. Two quick puffs get the butt burning, followed by one long drag.

I draw the smoke deep into my lungs, hold it for a bit, and let it dribble sinuously from my nostrils. In this fantasy, the air is still, since the smoke *must* be sinuous, as in an old Bogart[43] movie. I empty my lungs while eyeing the smoldering cigarette appreciatively. "We've been through a lot, old friend," Fantasy Me says. "What do you say, one last go and then we get on with things?" And for the next few minutes, I revel in the familiar motions of raising a cigarette to my lips, inhaling deeply, and exhaling again.

All too soon, as the stars twinkle into existence, I pinch off the head of the cigarette and toss the butt in a bucket. I open the front door and as I enter the foyer, my thankful and adoring children run to me for hugs. From across the living room, my wife beams. Gathering up the kids, I spare one last look at the porch and smile wistfully. Fade to black.

◆ ◆ ◆

[43] Dead at age 57 from smoking-related throat cancer.

In my previous 37 years of smoking, I had made one wholehearted attempt at quitting. In 1994, I went cold turkey for six entire weeks, mostly because the woman I liked at the time had very strong and negative opinions about cigarettes. My romantic overtures were not returned, however, and soon I was smoking again. Other than those few weeks, and a handful of half-hearted attempts, I was a happy pack-a-day smoker since I was 16 years old.

Of course, smoking was becoming less and less congruous with my new not-completely-unhealthy lifestyle. It was getting time to quit. The mere thought of which produced a strong and immediate sense of anxiety, the kind that ends most attempts to quit smoking before they even begin. In fact, my most recent half-hearted attempt, just four days earlier, ended only hours after it began, with my brain screaming at me to give it a goddamn cigarette already. I remember *that* day. I remember it mostly because I planned it in advance. It was *the day I was going to quit smoking.*

Putting that plan into action, I intentionally did not buy a new pack of cigarettes the day before, and on Saturday, July 8, I woke to just two remaining smokes. Following my normal routine, I went downstairs, had a cup of coffee, and then smoked what was meant to be my penultimate cigarette. Sixty minutes later, I smoked the last. It was awful. Not the cigarette, that was fine. What was awful was the knowledge that it was my last freakin' cigarette ever!

From the very moment I stubbed it out, regions all over my brain began to sing, and my stress levels ratcheted up. My mind knew that in about an hour it would be craving a smoke and the stupid body that was carrying it around wouldn't have any cigarettes. I envisioned tearing up the house, frantically looking for a forgotten pack. I wondered what the hell I was going to do after my next cup of coffee, after lunch, after dinner. How many people were to die that day? But, hey, it's the new me, right? I could do this.

I lasted seven hours. Around 4:00 that afternoon, rather than throttling the children, I walked down to the convenience store and bought a pack of cigarettes.

The Respiratory System

There are two principal pathways for turning glucose into energy: aerobically and anaerobically — with oxygen and without. If you need a small amount of energy in a hurry, you can quickly break down glucose anaerobically and call it a day. But if you want to get the biggest bang for your glucose buck, oxygen is absolutely essential to the process, and that means we need a system for delivering oxygen to all of the cells in your body. As luck would have it, we have just such a thing, the respiratory system.

Skipping all the bits above the shoulders — the nose, the larynx, etc. — the respiratory system starts in your neck with the trachea, which is nothing but a rigid tube. The trachea branches into two bronchi, right and left, which enter the lungs. The bronchi then branch into smaller tubes known as lobar bronchi. These branch again into bronchioles, which have different physiological characteristics than the bronchi. The bronchioles go on branching for a while, getting thinner and thinner. Honestly, it's all kind of boring. In humans this branching happens 23 times before finally terminating in the alveoli, where things get interesting again.

An alveolus (singular) is a microscopic air sac sitting at the end of this branching bronchial tree. In fact, at every terminal point, there are tens or hundreds of alveoli (plural), and some 500 million of them altogether. Each collection of alveoli is gently wrapped in a dense network of capillaries that have branched from the pulmonary artery of the heart. It's the alveoli's job to exchange gases, oxygen and carbon dioxide, between the atmosphere and the blood in these capillaries.

The air around us is 78% nitrogen, but even though nitrogen can be very reactive (think nitroglycerin), the nitrogen in the atmosphere is not. That's because the nitrogen atoms are bonded with each other to form a diatomic (two-atom) gas known as dinitrogen or N_2. Since this pairing happens via a rare triple electron bond, dinitrogen's outer electron shell is completely filled (5 valence electrons + 3 shared with another nitrogen), making N_2 very stable and largely inert.

Almost all of the rest of the atmosphere, about 21%, is made up of nitrogen's neighbor on the periodic table, oxygen (the missing 1% is hot air coming from Haskell developers telling you what a *real* type system looks like). Oxygen is an extremely reactive element, but it's made somewhat less temperamental by forming a double covalent bond with another oxygen atom, thus filling up its outer shell (6 + 2) and making O_2. It's this O_2 that's of interest to us, because it's what fuels our cells.

When we inhale, we fill our lungs with N_2 and O_2, as well as a little water vapor and carbon dioxide (CO_2). Our blood contains all of these gases as well, and in such proportions that the pressure of these gases in the blood is *nearly* the same as the pressure in the atmosphere. The arterial pressure of N_2, that is, its pressure inside the blood vessels, is 573 mmHg (millimeters of mercury, like in a barometer) and the alveolar pressure (which is the same as the atmospheric pressure) is also 573 mmHg. Since these two values are the same, no nitrogen is diffused across the respiratory membrane into or out of the bloodstream.

Things are a little different for oxygen. As you know, and as we'll discuss in more detail in Chapter 6, your cells pull oxygen out of the blood and use it to help power themselves. Eventually, the oxygen-poor blood makes it back to the heart and then to the lungs. But because there's now less arterial oxygen, there's a slight difference in pressure between the O_2 in the alveoli and the O_2 in the bloodstream. That is, a concentration gradient has formed between the two, causing the alveolar oxygen, the oxygen in the lungs, to diffuse across the blood/air barrier into the bloodstream, equalizing the pressure.

This oxygen doesn't just swim around in the blood plasma, the watery part of your blood in which all the useful bits are suspended. Instead, it gets picked up by your red blood cells, inside each of which are gathered about 250 million hemoglobin proteins. And inside each of these hemoglobin proteins are four iron atoms. The iron has a strong affinity for oxygen, allowing each hemoglobin protein to grab four oxygen atoms on its way past the alveoli. That's a billion oxygen atoms per blood cell. This now oxygen-saturated blood is directed back to the heart, where it is pumped with some force out to your muscles, brain, and other organs. Some seconds later, as the oxygenated blood flows by energy-starved cells, it will give up the oxygen in exchange for carbon dioxide.

As we'll see, when a cell needs energy for any reason — to move, divide, make proteins — it uses glucose (predominantly) and oxygen (usually) to create that energy. One of the byproducts of this reaction is carbon dioxide (CO_2), which is released back into the bloodstream. When a red blood cell swims through a cloud of CO_2 that marks a region where oxygen is being used up, it is induced to give up the oxygen it's carrying.

This happens because some of the CO_2 floating around in the blood plasma is taken inside the red blood cell. Of that CO_2, a small amount may bind with the hemoglobin and ride it back to the lungs, but the bulk of it is set upon by an enzyme that lies in wait, carbonic anhydrase. Carbonic anhydrase joins the CO_2 with some H_2O to make carbonic acid, H_2CO_3. As it turns out, carbonic acid is pretty unstable and quickly breaks down into one positively charged hydrogen atom (aka H^+, aka a proton) and HCO_3^-, aka bicarbonate. This negatively charged bicarbonate is ejected back into the plasma (in exchange for a chloride atom), where it remains until it gets back to the lungs. In the meantime, that H^+ needs to be dealt with so it doesn't raise the acidity level of the blood (because that's what acidity is, remember, an excess of protons in a solution). Fortunately, the H^+ will react with the hemoglobin, and that — finally — triggers the release of the oxygen so it can be picked up by the nearby cells.

In short, red blood cells convert some of the waste product of cellular energy production, carbon dioxide, into Alka Seltzer, and in so doing release their oxygen, giving it to the cells that just used up theirs and thus need it most. This bicarbonate, by the way, actually does play a role similar to Alka Seltzer; it keeps the blood's pH level, its acidity, constant.

This oxygen-poor — or if you will, carbon dioxide-rich — blood swims through the remaining capillaries, into the veins, and back to the heart, which gives it a push, sending it on to the lungs. But before it can pick up more O_2, it needs to rid itself of the remaining CO_2. Fortunately, the concentration gradient for CO_2 is running the other way, as there's more CO_2 in the blood than there is in the alveoli. Since most of the waste CO_2 is dissolved in the blood plasma, it will simply ride the concentration gradient in the reverse direction of the oxygen, into the alveoli, where it is exhaled into the atmosphere so that plants can turn it back into oxygen and we can start the whole cycle over again. All very neat. Also at this time, the CO_2 that was turned into bicarbonate will slide back into the red blood cell, where quick as a flash, the same carbonic anhydrase enzyme knits it back together with its missing proton, making CO_2 and H_2O again, which also get exhaled.

Now let's step out into the rain and spark one up.

Thank You, Captain Obvious

Google "cigarettes" and you will find hundreds of sites that discuss why they are unhealthy. Most of them will begin by saying that cigarettes have over 4,000 chemicals in them. I have no idea where that number comes from. Nor do I know if those chemicals include dihydrogen oxide (water) or sodium chloride (salt), or whether all of those 4,000 chemicals are present in an unlit cigarette or only in the smoke, or how high the concentrations are, and so on. Regardless, the list of scary substances in cigarettes and cigarette smoke is truly sphincter-tightening. These include benzene, polonium-210, arsenic, cyanide, and more. Many of them are naturally found in soil (and therefore plants), the rest being fertilizers, pesticides, processing agents, and additives.

It's painfully obvious that repeatedly burning these chemicals and drawing them into our lungs is damaging to our health. That said, we don't have to link any individual chemical directly to the health condition it may be responsible for. We'll never be able to do that (though the polonium and a radioactive form of lead, lead-210, are extremely suggestive of cancer). The fact is, it doesn't matter much whether you're smoking tobacco, cannabis[44], or kale. What you're really doing is oxidizing complex carbon and hydrogen molecules at 1,600° F, and that's going to generate an aerosol (the smoke) full of all sorts of nasty-ass hydrocarbon particles, e.g. tar and benzene, adulterated with whatever the manufacturer felt like adding to make the experience more palatable.

Take a drag on that cigarette and all that particulate matter travels the length of your respiratory system, most of it twice (in and out): carbon monoxide, cadmium, ammonia, all of them painting your mouth, nose, bronchioles, and alveoli black with tar and eliciting a protective, but ultimately harmful, inflammatory response. This inflammation leads to a restriction in the amount of gases that can be exchanged through the alveoli. Provoke that response often enough and it will become persistent. Then you're on your way to bronchitis or, worse, emphysema, both of these collectively known as chronic obstructive pulmonary disease, COPD.

But don't worry, I'm not going to dig into COPD or lung cancer or heart disease or a smoker's increased risk of infection or anything else on the staggeringly long list of adverse health conditions linked to smoking. It just doesn't seem illuminating to spell out the side effects of smoking 270,000 cigarettes. You know and I know that smoking will fuck your shit up, and you hardly need a background in statistics to know it's directly linked to most cardiopulmonary diseases. In fact, smoking is responsible for approximately 6,000,000 deaths every year. In the U.S. alone, the medical costs related to smoking exceed $150,000,000,000.

Obviously, smoking a single cigarette is not going to kill you (although it may if you are asthmatic or have certain allergies). As mentioned, in most people (though not me) it will elicit an inflammatory response, maybe a coughing fit, but that will subside quickly and the body will repair what little damage may have been done. If that were it, we wouldn't be discussing smoking, because nobody would be smoking. But there's one more thing smoking does. One more chemical in that miasma of toxins that we need to discuss. It's time to talk about addiction.

---

[44] While I fully support the legalization of marijuana for medical and recreational use, don't think for a minute that it's healthier than tobacco; it's not, though it is unhealthy in slightly different ways. Similarly, for some, vaping *may* offer a path toward quitting, but I wouldn't switch to it for any other health reasons — lord knows what's in the nicotine solution.

Neurochemistry, Part I

As a human, you are very special, it's true. Unique among the animals of Earth, you can reason abstractly, ponder the future, build complex tools, and promulgate Russian propaganda on Facebook. What a piece of work you are! From a biological point of view, however, you're not all *that* unique. By almost any measure, you're not very different from a dog or a dolphin (though you are almost certainly less pleasant). You are a replicator, descended from a very long line of replicators. You are, frankly, little more than a big bag of carbon programmed to fuck.

That programming, of course, takes place in your brain. Your brain is itself just a collection of highly specialized carbon-based molecules, aka cells. It's these cells that make you infatuated, restless, curious, aroused, guilty, happy, calm, and basically everything. Since there's no magical homunculus living inside your skull, all of these mental states are nothing more than manifestations of the brain generating and/or reacting to chemicals (and some electricity).

These specialized cells are called neurons, and you have around 90-100 billion of them. There are some other supporting cells (glial cells) in your brain, almost as many as there are neurons, but it's the neurons, also known as nerve cells, that make you *you*. There are a variety of different types of nerve cells: motor neurons (attached to your muscles and telling them to move), sensory neurons (picking up signals from your eyes and ears, etc.), and interneurons (doing all the readin', writin', and 'rithmatic). It's this last group that's interesting to us.

These interneurons, typically referred to as just "neurons," since they are of primary interest to everyone else too, are generally made up of three important components. There's the "soma," which is pretty much what you think of when you think of a cell. Branching away from the soma are the brainy bits: the dendrites, a dense tree of filaments that are the receivers of chemical and electrical information; and the axon, a single branch reaching out the other way that transmits many of these chemical and electrical signals. The axon can stretch a considerable distance in order to be in transmitting range of the receiving neuron's dendrites. In the case of your sciatic nerve, the axon may be as much as three or four feet long. The neurons in the brain are densely packed, but the axons of one neuron do not actually touch the dendrites of another. Instead there is a small gap in between called the synaptic cleft. Collectively, the transmitting end of the axon, the receiving end of the dendrites, and the cleft in between are known as a synapse.

Across these synapses flow neurotransmitters. Scientists have identified more than 100 different neurotransmitters of varying chemistry. Most neurotransmitters are either free amino acids, peptides, or monoamines (an amino group linked to a simple carbon chain). Neurotransmitters are typically manufactured in the soma, then stored in the axons, and under the right circumstances, they will be released across the synaptic cleft to specialized receptors on the neighboring dendrites. Of course, it's not just one neuron signaling one other neuron. It's typically a cascade of signals across a cluster of neurons which form those chemical exchanges you know as joy, anger, fear, contentment, and desire.

There's a very specific chemical behind that last emotion, desire, it's a neurotransmitter called dopamine. Dopamine is a monoamine, and when it binds to a receptor, your brain interprets that as pleasure. Dopamine is part of our reward system, which was laid down by evolution over the eons to encourage us to do things that make us successful replicators. Things like eating, having sex, and investing in our offspring. Do rewarding things and you'll get a little buzz of pleasure. These rewarding things are not just intrinsic (food and sex), they can be extrinsic as well (money and power), since we have learned to associate these gains with the eventual reward we can use them to attain. Indeed, the reward system can be triggered by many causally related events: the smell of onions sautéing in butter, walking past a tavern, or viewing pornography. It is a fundamental system inclusive of all things that motivate us to live and thrive. Take away the dopamine and most of us wouldn't even get out of bed in the morning.

Our reward system can be broken into two closely linked steps: wanting and liking, sometimes referred to as seeking and consummating. That is, evolution has primed us to both "want" pleasure and to "enjoy" pleasure when we get it. The wanting is what motivates us. "I want a milkshake." The liking is the payout. "That's a pretty fucking good milkshake." The wanting part of the reward system, the goal-oriented part that keeps us striving and moving forward, is the more prominent of the two, and is triggered primarily by the release and reception of dopamine in various parts of the brain. In particular, the dopamine projection neurons are located in the ventral tegmental area (VTA) of the lower midbrain, and the dopamine receptors are in the nucleus accumbens. If something, anything, can cause the dopamine receptors on the nucleus accumbens to be stimulated, you are most definitely going to want that something again. You will *crave* that something.

The liking (consummating) part of the reward system is physically smaller. There are tiny clusters of opioid and cannabinoid receptors scattered around the brain. These are the so-called "hedonic hotspots," more commonly known as the pleasure centers of the brain, even though "hedonic hotspots" is clearly the superior name. These hotspots are, as I'm sure you have guessed, stimulated by the body's self-manufactured class-one drugs, such as endorphins (an opioid) and endocannabinoids.

Although the seeking and consummating parts of the reward system are frequently stimulated in rapid succession, if not simultaneously, it is not necessary to trigger any of the brain's hedonic hotspots to get a jolt from dopamine. Neither cocaine or gambling, for instance, activate the brain's pleasure centers, but dopamine release is through the roof. In the typical ritual of flirting, dating, sex, and orgasm, only the last is not dopamine driven. And relative to the buildup, consummation is brief and occasionally disappointing. The pithy way of putting this is, "It's not the kill, it's the thrill of the chase."

All of this was wired into our brains long ago to make sure we got enough calories, ran away from tigers (not dying is also rewarding), and sought out a mate. Unfortunately, in modern society, there are a lot of substances and behaviors that can stimulate, indeed overstimulate, the reward system, even as many of the dangers of the past (starvation, tigers) are no longer quite so prevalent. One of these substances is nicotine (others include sugar and alcohol).

Your brain also manufactures a neurotransmitter called acetylcholine (ACh). It has a number of roles in the body, but when it's picked up by receptors in the brain, known as "nicotinic receptors," its principal job is to accentuate learning and memory, partly by making you more alert and energized. ACh also activates a dopamine response, which, of course, triggers a buzz of pleasure. As you have probably guessed from the name, nicotinic receptors don't solely respond to ACh; they will happily bind with nicotine.

Ordinarily, your brain regulates how much ACh is released and where to release it to, and it is very parsimonious when it does so. However, approximately ten seconds after your first puff on a cigarette, your brain is positively swimming in nicotine. All this extra stimulation makes the smoker very focused and very happy (and more likely to remember the stimulus). But wait, there's more. Nicotine also stimulates your brain to start spewing other neurotransmitters, including endorphins. Endorphins are interesting for a variety of reasons. Important here, and as just discussed, is that they bind with opioid receptors in your brain's hedonic hotspots, just like heroin does, and provide a brief sense of euphoria. So, yeah, double whammy there. Nicotine triggers both halves of the reward system: seeking and consummating. It's like getting laid on the first date.

In addition, nicotine triggers the release of both serotonin, the body's natural happiness maker, and adrenaline, the body's natural stimulant. This combination of neurotransmitters is what gives nicotine its unique property of being both a stimulant and a sedative at the exact same time. Is it any wonder I miss it so? It makes me wonder why the tobacco industry never got together with the food industry (after all, they're cut from the same cloth) and introduced nicotine into a variety of products, the way we find caffeine in Red Bull, Mountain Dew, and Hydroxycut. I mean, all by itself, and in reasonable doses, nicotine is not poisonous. It's not even directly implicated in any of the negative health consequences of smoking.

One last thing before we temporarily leave neurochemistry behind, and that's the need for more and more. A healthy adult is releasing some amount of dopamine, beta-endorphins, ACh, and so on all the time, and the corresponding receptors are continuously being tickled. But in the throes of addiction, as you artificially drive up the release of dopamine and endorphins, the corresponding receptors begin to fail, leading to the desire, in my case, for one more smoke, one more cookie, or one more drink in order to satisfy my cravings. Overstimulate these receptor neurons too much, and they start to die, necessitating an even higher level of substance abuse to attain the same level of pleasure.

Smoke! Smoke! Smoke that Cigarette

Given the declining prevalence of smoking in the U.S. today, it's very likely that you are not a smoker. Despite the quick science lesson you just got on addiction, as a non-smoker, you likely have only a syllogistic grasp of what nicotine addiction is all about: cigarettes contain nicotine, nicotine is addictive, therefore cigarettes are addictive. If you have your own addictions, maybe you're translating in your head. If you have no addictions of your own, maybe you can intellectually understand that addiction is at least something that exists[45]. Regardless, it's very likely that you don't fully understand the appeal of smoking. Or rather, what used to be the appeal. I swear, the smartest thing you sanctimonious, non-smoking, sons o' bitches ever did was kick us smokers out into the cold.

In 1987, the year of my first real job out of school, I was able to smoke at my desk. and it was a wonderful thing. When you were deep in thought trying to puzzle out why a Wyse-50 terminal in VT-100 emulation mode was sending the wrong escape sequences over the serial line (it was a long time ago), it was so nice to just light up and ponder until the answer simply and inevitably revealed itself. The CEO absolutely despised smoking, but there was little he could do about it, since everyone else with a "C" in his (there were no "hers") title was a smoker.

---

[45] If you think you're addiction free, you may want to think again. Addictions are not just about consuming psychoactive drugs such as alcohol, nicotine, or heroin. You can also be addicted to exercise, pain, thrill seeking, gambling, sugar, sex, pornography, caffeine, the internet, and much more. In all cases, the reward system is engaged.

That same year, Philip Morris made $61 billion in inflation-adjusted revenue and 28.8% of Americans were smokers.

Five years later I switched employers. At my new company, while you could still smoke in the building, you could no longer smoke at your desk. Instead, we smokers had to huddle in the hallway leading to the men's room. Ignominious, yes, but at least we were warm and dry. And when work was over, you could still go downstairs to the Chinese restaurant with the Irish bartenders and have a smoke there.

The restaurants and bars in New York were smoker-friendly in 1991, which is not surprising given that fully one quarter of my fellow Americans still smoked. I have to say to you non-smokers, the ability to stool up at a bar, order an IPA — wait, this is 1991, order a Michelob — and light one up was pretty nice. And when you ran out of smokes, you could just saunter over to the vending machine, plug in eight quarters, and help send Philip Morris's inflation-adjusted income to $100 billion.

A year later, my company offered me an opportunity in Chicago, and I took it. We leased space at 900 N. Michigan Ave, and when I got there the entire building was non-smoking. Smoking outside during those Chicago winters could be pretty brutal, but the bars were still smoker-friendly and so too, of course, was my apartment.

And that was the best place of all to smoke. Picture this: it's, let's say, a Tuesday evening in September. There's not much going on, no Internet to speak of, and nothing on TV. You pour yourself a glass of wine, pull a Wodehouse novel from the shelf, lay down on the sofa, and puff enjoyably along with the good eggs at the Drones Club.

Bear with me here, non-smokers (and non-Wodehouse fans), this is the part you don't fully understand yet: the ceremonial, lifestyle aspect of smoking. I'm speaking of a behavior so integrated into your life, into your very sense of self, that just going through the motions is a rewarding experience. Smoking is cool! Smoking is sexy, too. And fun, and mysterious, and edgy, and agreeable. While smoking, I can assume a contemplative air. I can be affable or dangerous or focused or worldly. Cigarettes are both friends and friend-makers. Accepting the offer of a cigarette from a stranger creates an immediate bond. Offering one to someone you're attracted to is flirtatious. Alone in my apartment, I can commiserate with Bingo Little, light the newly rolled cigarette of Brigid O'Shaughnessy, even sit in the shadows of an inn in Bree and wonder what those fool Hobbits will do next.

If none of that resonates with you, picture yourself ordering and enjoying a fine Scotch at Bemelmans; or a silky, black coffee at a cafe on the Croatian coast. Imagine browsing a used bookstore in London, or driving a roadster through the Santa Ynez mountains. If a frisson ran down your back thinking about any of that, you have a better understanding of what it meant for me to be a smoker. It wasn't just the chemical dependency, cigarettes were alluring, comforting, and refined. They were a significant part of what it meant to be me. Yes, the tide had to turn, smoking is terrible and I had to quit, but its absence has left a hole in my life larger than you can ever imagine.

The afternoon of November 13, 1998 (Philip Morris revenue: $111 billion), was the last time I lingered over a cigarette. I've smoked plenty since then, of course, but I haven't been able to linger over them. I know this, because later that evening I was married, and I haven't smoked inside since. Rain or shine, sleet or hail, windy or still, I spent the next 19 years huddled in the lee of a doorway, smoking as quickly as possible (two and a half minutes, if you're curious) simply to satisfy my addiction, and enjoying it not a bit. I should have quit then. I remember trying to wean myself off cigarettes early in my marriage, but it was one of the half-hearted attempts and didn't take.

In 1995 California passed its statewide smoking ban, prohibiting smoking in all enclosed workplaces, including bars and restaurants. At the time, 24.7% of Americans were smokers. In 2003, New York did the same, followed by Massachusetts in 2004. By then, 21.6% of Americans continued to smoke. In 2008, Illinois joined the parade. Smoking declined to 20.6%. In 2017, the most recent year for which data is available, 14.0% of Americans smoked. This is down from a stunning 42.4% in 1965 and 33.2% in that eventful summer of 1980.

I'm sure statisticians can pluck apart the data to determine what precipitated this dramatic decline in smoking. Maybe it's awareness of the health effects? In 1964 the Surgeon General released the first report linking smoking with lung cancer. Since then, the glaringly obvious link between smoking and cancer, pulmonary disease, emphysema, gum disease, and frankly damn near everything else has only gotten stronger. But honestly, we always knew that smoking will kill you. They were calling cigarettes "coffin nails" long before 1964. Nevertheless, the tobacco industry will always bear the shame of their spin campaign vilifying science, a technique now being put to effective use by climate change deniers and the processed food industry.

Or maybe it's the cost that's driving down smoking? In 1998, the four major American tobacco companies famously settled litigation with 46 states to the tune of $206 billion, enforced and payable over the next 25 years[46]. That was the first time, though not the last, that the tobacco industry failed to prevail in legal action. The impact of that settlement, as well as a host of federal, state, and local taxes, has sent the price of cigarettes soaring. In the 37 years that I smoked, the cost of a pack of smokes has outpaced inflation by 600%. As I told you, I was buying cigarettes in Brooklyn for $0.75 a pack ($2.22 adjusted) in 1980. That same pack costs around $15 today ($13 officially, but I haven't seen that price in NY in years).

Or maybe it's that smoking is now taboo? The social censure campaign has been wildly successful. Not only have smokers literally been kicked to the curb, but no one even has to feign politeness anymore. Smokers are sad, uneducated, revolting people, who don't need to be tolerated or even acknowledged.

True story: one afternoon a few years ago, I was standing outside a restaurant, smoking. A little girl of about five came out of the restaurant with her mother. The girl saw me take a drag and exhale, then she started to actually scream. Not in the almost-cute way of not understanding how smoke was coming from a grown man's nose, but because I WAS SMOKING! By 2025, I predict people will be hunting smokers for sport, and that little girl will be up in a blind with her trusty bow, watching. Always watching.

To one degree or another, the decline in smoking can be contributed to all of these things: the health effects, the money, the ostracization. But from the viewpoint of this ex-smoker, the primary driver is that you took all the enjoyment, the ceremony, the cool out of it. Still, I persevered. For the last 19 years of marriage, if I wanted a smoke, I got up off my own couch and went out into the elements. Right up until I didn't.

Cold Turkey

Like most Americans, I live within a few hundred feet of a CVS. On the morning of July 12, 2017, after my three-mile walk and having gotten ready for work, when I would normally have had my first cigarette of the day, I walked over to the CVS instead. I did not *plan* to quit that particular Wednesday. I went to bed a smoker and expected to wake up a smoker. But there I was asking the cashier for a box of nicotine patches (they're kept behind the counter). Standing off to the side of the registers, I read up on the basics of the patch.

---

[46] Are you paying attention, Coke and Pepsi? This is your future.

It's an eight-week program. For the first month, you wear a daily patch that delivers 21 mg of nicotine into your bloodstream, about the same as you would get from smoking a pack of cigarettes in a day. Then you step down to 14 mg for the next two weeks. Then 7 mg for the last two. Then nothing.

I bought a CVS brand 7-day pack of Step 1 patches and an 8-pack of Breath Savers mints. Back at home, I slapped a patch on my left shoulder, put a mint in my mouth, and drove to work. At work, I murdered the guy in the neighboring cube for typing too loudly. That's not true. I was able to keep it together. Every hour, when I would normally get up and have a cigarette, I got up and had a mint instead, while I walked around the parking lot to get my steps.

That evening, Denise, Henry, and I went out to dinner (Ambrose was away at camp). After dinner we walked home. No one remarked on the fact that I had not smoked in the last three hours that they'd been around me. Not when I got home from work, not on the walk to the restaurant, not after the appetizers (which I didn't have, since I was still dieting), not after (¾ of) the main course, and not on the walk home.

"Notice anything?" I asked.

Denise and Henry looked around.

"About me?"

"Yes, honey, you're losing weight. We get it," said Denise.

"Not that. Anything else?"

She looked me over. "We give. What?"

I held out my hands, fingers splayed. They looked at me quizzically.

"No cigarette." I said. "I haven't smoked all day."

"What? Really? That's fantastic!" said Denise. "Good for you." Then she paused, narrowed her eyes, and spat out a question. "What's my favorite book?"

"Pride and Prejudice."

"What performer do you hate?"

"Stevie Nicks."

"What was I wearing on our first date?"

"I have absolutely no idea."

She seemed to consider things for a moment. Then she leaned in close, her lips next to my ear. "I'll accept for now that you are who you say you are, but I've got my eye on you."

I made it through the rest of the day without violence. The next day set the tempo for the next few weeks. Wake up and do some kind of cardio, shower, shave, apply patch, dress, eat a healthy breakfast, and go to work. At work, when I would normally go outside for a smoke, I'd pop in a mint and take a walk around the parking lot instead. If the weather was unpleasant, I would just go up and down the stairs and hallways.

This time quitting took. As I write this in late July 2018, I've been smoke free for over a year. Thinking back on what was different this time, a few things stand out.

One, I don't think I could have done this without the aid of nicotine patches. If you're a smoker thinking of quitting, I would highly recommend you use the patch too and don't try and wean yourself off cigarettes unassisted. Cigarettes are exceptionally difficult to quit. More so than nearly any other addictive substance or practice, and just as difficult as quitting heroin or cocaine. While I know people who have toughed through the withdrawal symptoms, I don't think many can. I know I couldn't.

Two, having a physical substitute for a cigarette is helpful; in my case, mints.

Three, quitting without planning to quit was HUGE. It eliminated all of the anxiety that keeps you from even trying to quit in the first place. So, dear smoker, should you wake up one morning and find that you don't have any cigarettes left, instead of going to the 7-Eleven, go to the CVS that's inevitably right next to it, and pick up some nicotine patches.

Fourth, I think exercise has a lot to do with it. Not just because it's hard to smoke while you're jogging, but because exercise makes for a great nicotine replacement. As we'll discuss in more depth in Chapter 8, exercise lowers stress levels and the accompanying cortisol release that might drive you to reach for a cigarette. And it stimulates the release of dopamine, endorphins, and serotonin, just like smoking does.

Finally, and most importantly, it was just time.

This is hard to express. I did not finally come to my senses and weigh the health benefits of quitting more highly than the pleasures of smoking. Remember, I and every other smoker are intimately familiar with the toll of smoking, and we always have been, disinformation campaign notwithstanding. We are unwise, we are not uninformed. Nor was I suddenly blessed with an excess of wisdom. Also, it's not simply that all the fun was gone. Sure, throwing us smokers out in the cold really does kill the allure of smoking, but that's been underway for 20 years. In this regard, there was nothing special about July 2017.

Honestly, I think what made that moment different from any other is that I wasn't a smoker anymore. That is, I wasn't — didn't want to be — the kind of guy that smokes. Instead, I enjoyed exercising, and thus being someone who exercised. I enjoyed being slender again, and thus someone who took care of himself. I liked this new me. My wife liked the new me. The new me was more fun, more interesting, and more healthy. I didn't want to be the old me anymore. I wasn't the old me anymore. I had successfully rewired my reward system. (This is also why exercise is Dieting Rule No. 3.)

Of course, the same was true three days earlier, when I tried and failed to quit. That's where all the stuff about nicotine patches and mints comes in. Even if, like me, you no longer think of yourself as a smoker, you're still addicted to nicotine and you will still want a cigarette. Hell, it's been a year since my last smoke, and while it's getting easier, I desire a cigarette all the damn time!

Don't imagine quitting is easy. It's not. Trust me, it's not willpower that's going to get you through it. Few people have the willpower to resist the ever-present call to smoke, day after day, for months and years. Willpower alone might get you through a week, and then you'll fold like a smoker from our dystopian future who's just taken an arrow to the gut. Quitting smoking is all about redefining yourself. I don't smoke, not because I don't want to smoke, but because I am not a smoker.

I strongly suspect that the recidivism rate for ex-smokers is so high — 90% resume smoking within eight days — simply because reason (it's healthy) and comfort (I'm warm and dry) and strength of will (I can do this) are not enough to overcome the addiction. But that's all most people bring to the table. I contend that you need one more thing: smoking can no longer define you. My path — my completely unexpected, unplanned, quite possibly unreproducible path — toward that state of mind was exercise and diet.

Will I ever smoke again? Given the statistics, I'd be foolish to say never, but I don't think so. I'm pretty sure I'm done. It was a good 37 years, and they will probably catch up with me down the line. But no, I likely won't ever smoke again. I'll tell you this, though, should I ever get a "you have six months to live" diagnosis, I might just spend those six months wrapped in the blue-gray haze of cigarette smoke.

◆ ◆ ◆

And now a confession.

That last cigarette I discussed back at the beginning of this chapter, the one I can't remember, it wasn't really my last cigarette. My last cigarette was two weeks later. That previously discussed cigarette was the more important one, and I legitimately cannot remember it, but it wasn't the last.

Around the end of July, my wife and kids were away and I was a bachelor for a weekend. It was a Friday morning and I was fishing around for something in the excessively pocketed backpack I use to carry my laptops (plural) around. As I rummaged through old boarding passes, loose change, and a rat's nest of USB cables, I came at last to a little-used pocket containing — what else? — a long-forgotten and unopened pack of cigarettes. A better man than I am would have tossed them straight in the trash. I zipped the pocket closed again.

At work, that bag sat at my feet all day. Cartoonishly, the cigarettes inside the bag gnawed at my conscience. "Just one," they said. "It won't kill ya. We mean, sure, over time it'll kill ya, but that's years from now. Besides, all the cool kids are … Well, they're not smoking anymore. We don't know what the cool kids are doing these days. maybe they're sipping gluten-free cocktails or vaping or something. But that's neither here nor there. Just have a smoke. You know you want to." Chatty little SOBs they were.

Throughout the day, every time I wanted a smoke, it really was an act of will not to just grab the pack and go see the old gang outside. (See, you have to save your willpower for when you really need it.) Of course, I should have tossed them out, but I didn't. Instead I made a plan, a plan that involved no willpower whatsoever.

There are certain foods I really like that Denise doesn't: steak, liver, corned beef, and so on, and we don't typically have them for dinner. So, when Denise is away, I make something from the "do not serve to family" list. My plan, then, was to buy a bright crisp rosé and a big chewy ribeye along with lima beans, couscous, and caramelized onions (I was done with the diet by then). Then, after a wonderful meal, I would sit on the porch and have one — just one — last cigarette. To actually live the fantasy described earlier.

I ate my steak. I drank my wine. I went out on the porch. I plucked a cigarette from the pack, ran it under my nose like a fine cigar, and placed it between my lips. In one fluid motion born of long practice, a lighter materialized. A flame was lit and touched to the business end of the cigarette. Two quick tokes got the butt lit, followed by one long drag. I drew the smoke deep into my lungs, held it for a bit, and let it dribble sinuously from my nostrils. Exactly as envisioned.

Hmm. That wasn't nearly as nice as expected.

I took another drag and waited for that buzzing sensation you get when it's been too long between smokes. It didn't come. I was still wearing the patch.

This sucks.

I smoked the cigarette down to the filter. It was deeply unsatisfying. It didn't even taste good.

Well, that didn't go at all as expected. I suppose it's all for the best, though.

I walked into the kitchen, ran the rest of the pack under the tap, refilled my wine glass, and watched a guy movie. A Denzel Washington flick this time. *Training Day*, I think.

Chapter 5: Grow

My siblings and I don't much resemble one another, but even so, it's not too hard to separate us into my father's side of the family and my mother's side, into Laceys and Fitzpatricks. I — more so than my brother or any of my four sisters — am a Lacey. I am quite obviously my father's son.

I have a picture of him when he was about my current age. It's a simple snapshot. He's sitting on our stoop (in Brooklyn we have stoops) wearing paint-splattered shorts and an undershirt, and reading a book. With his left hand, he's pushing wavy salt and pepper hair back off his brow, with the other he's holding a can of Budweiser, the book is balanced on his lap. His undershirt reveals what I remember as most prominent about him physically: his muscles. He did not have a farmer's build or anything even close. In fact, if you were to use a single word to describe him, it might be the same used to describe my younger self, "lanky." His biceps, though, were always pronounced, so maybe "sinewy" would be the more appropriate adjective.

Late in June, the boot camp class officially came to a close. My thrice-a-week exercise program was over, and I needed a replacement. I switched my Tuesday and Thursday walks to Monday, Wednesday, and Friday, thinking that was enough cardio (it wasn't, as we'll see in Chapter 7). But I also needed to continue to tone the muscles. The weight loss was starting to become evident, and the exercise to date was beginning to add some definition to my paternally-gifted biceps. I wanted to capitalize on that.

I decided I would double down on strength training. Actually, that's not quite true. At this time, I had no idea that strength training was even a thing. In the sense of it being good for you, that is. I didn't want to get healthy via weight lifting, I wanted to make those Lacey genes do their thing — I wanted to get jacked. And that's when I learned that the new me was also vain.

This is novel. I do not consider myself a handsome man, my clothes are largely from Marshalls, and my hair is "styled" at Supercuts. No one has ever accused me of being vain. Arrogant? Sure. Opinionated? Absolutely. Derisive? On occasion. But never vain. Yet the one thing I wanted most out of this whole transition were some guns. That I had no idea how to go about getting them was of little concern.

I had no idea how to build muscle — and this should come as no surprise to you by now — because I had no idea I would be building muscle. Six weeks earlier, when all this began, there was no plan to be ripped, swole, or even buff. Back then, there was only the notion of exercising enough to stop the hurt, which by this time was largely accomplished. To be healthy, all I really needed to do was continue the light muscle-toning exercises I'd been doing at boot camp. But in for a penny, in for a pound. On top of the healthy diet and weight loss, the smoking cessation and the cardio, I needed to lift weights.

Muscle and Bone

There are three types of muscle tissue to be found in the human body: smooth, cardiac, and skeletal. Smooth muscles are the ones that push things around inside your body, things like blood, food, and little baby humans. Smooth muscles are not controlled voluntarily, but instead are driven by signals from your autonomic nervous system. Because of this, smooth muscle tissue is not the subject of this chapter. Cardiac muscle, found, obviously, in your heart, also contracts involuntarily, and will be discussed briefly in Chapter 7. Skeletal muscle, however, that's the star of the show. Skeletal muscle is attached to bones and is under your conscious control. You use it to walk around, cock your head, hold a frying pan, chew, push a vacuum, dance, stand up, throw a ball, climb a fence, step on the gas, and, for those particularly blessed, raise one eyebrow and wiggle your ears.

To complicate matters, skeletal muscle comes in two distinct types: slow twitch (type I) and fast twitch (type II), and they're very different from each other. In general, slow twitch muscle tissue is used for tasks that require endurance, such as walking, jogging, and cycling. That is, tasks that are typically done slowly over long periods of time. Slow twitch muscle does not fatigue quickly. Indeed, as any long-distance runner can tell you, slow twitch muscles can go for hours and hours without tiring. In doing so, however, these muscles consume great gobs of fuel and oxygen. Another way of putting this is that slow twitch, type I, muscle is the kind you call on when doing cardiovascular exercise.

Fast twitch muscle, on the other hand, is for tasks requiring strength or speed, such as moving furniture, sprinting, or punching Nazis. Fast twitch muscle can deliver a lot of force very quickly, but it tires just as quickly, which is why even Olympic sprinters begin to fatigue around the 300-meter mark of the 400-meter event. When compared with slow twitch muscle, fast twitch muscle fibers use considerably less oxygen. In fact, fast twitch muscle is predominantly anaerobic[47].

All of this talk about jogging and sprinting may lead you to believe that one or the other muscle type is in use at any one time. It's not. In the real world, you generally call on both to complete a task. In fact, most athletic events require both endurance and strength: basketball, baseball, even the 800-meter race. Also, don't be under the impression that fast twitch muscles are over here and slow twitch are over there. In any particular muscle, such as your gastrocnemius (calf), these two types of muscle fibers are mixed and mingled alongside each other. How much of either type a person has varies based on genetics, age, and lifestyle, but on average, human skeletal muscle tends to be a 50/50 mix. Athletes, however, can have much more or less of either type. Marathoners, for instance, may have up to 70% slow twitch muscle fiber, while sprinters may be exactly the opposite, with 70% fast twitch. Furthermore, for us mere mortals, different groups of muscles have different percentages of fiber types. The muscles in your legs tend to contain somewhat more slow twitch fibers, while your upper body muscles tilt toward fast twitch fibers.

Although I didn't know it when I set out to build muscle, strength training is as important as cardio to good health, especially for those of us who are awash in AARP mailers. As you age, you naturally lose muscle mass. This deleterious process begins as early as age 30, but doesn't really get rockin' until you have 50 in your rearview mirror. At this point there's a gradual reduction in the number of muscle fibers (cells) as well as the size of those fibers. If these muscle fibers get too small, they will simply die. This loss in muscle mass, not unexpectedly, results in a corresponding loss in strength. A typical sedentary male is 15% weaker every decade than the decade before. In addition, the risk of injury increases while the rate of recovery decreases.

That's not insignificant, 45% of an adult male's body mass is muscle (37% for females), and 50% of that can be lost by the time you're 80. The consequences of this are profound. Sarcopenia, the medical term for age-related muscle loss, is implicated in any number of adverse health conditions including insulin resistance, rheumatoid arthritis, and general frailty. But muscle loss, i.e., atrophy, is not distributed evenly among the various muscle types. Since you need your

[47] To be complete, fast twitch muscle tissue is further subdivided into type IIa and type IIb, but the difference is subtle and not important here.

type I, slow twitch, muscles just to get from the couch to the bathroom, you tend to lose those more slowly. But if you don't use your type II, fast twitch, muscles, they will start to atrophy very quickly. Now here's the kicker: if you follow a reasonable cardio plan and go for, say, a three-mile run a few times a week, you are mostly *not* using your fast twitch muscle fibers, and they will *still* atrophy. You need to strength train.

Bone density follows a similar trajectory. Up until you're 30 or so, you are making as much or slightly more bone as you are losing. After that, the scale starts going the other way, though only gradually. Around age 50, however, a big cosmic thumb comes down on the wrong side of that scale, and both men and women begin to lose bone density at an alarming rate. Setting aside increased bone resorption due to menopause, men and women lose bone at roughly the same rate. Unfortunately, you can't set aside menopause, because the rate of loss is too significant. In other words, women get the short end of the stick on this one, though men are not immune.

Bone density varies from scapula to hip to femur, but it declines everywhere. With that decline comes the all too common fractured hip, bent spine, and awkward gait. Bone resorption has many causes: genetics, hormones, and diet among them, and, as usual, smoking and drinking. Lack of physical activity does not directly *cause* osteopenia (the halfway point between healthy and osteoporosis), but it accelerates it.

Similarly, there is no single cause for muscle loss. But unlike osteopenia, sarcopenia has one causative factor that outweighs all others: inactivity. Some decrease in an older person's level of activity is expected. After all, you can't fend off the ravages of time forever. There's probably a point in the not too distant future when even Tom Brady will retire. But you can absolutely give time a run for its money. With simple strength training — and it has to be strength training — you can stave off the onset and impact of sarcopenia for a very long time.

It doesn't take much to start tapping the brakes on muscle loss; a few push-ups and squats is a really good start. Or grab a pair of five-pound dumbbells and do a few curls and some lunges. Just get those muscles contracting, the more muscles and the more contracting the better. The muscle use will spur muscle growth and, conveniently, the muscle growth will spur bone growth. At some point you may consider adding resistance in the form of heavier weights or bands, but it's not required. Regardless, regular strength training will not only slow down muscle loss, it will reverse it. You will add muscle.

With the increase in muscle and bone comes a decrease in the incidence and intensity of ailments. For one thing, strength training will significantly improve insulin sensitivity (so will cardio). If you have type 2 diabetes or you're prediabetic, improving your sensitivity to insulin will reduce insulin swings and the resulting changes in body weight. The same process that boosts insulin sensitivity, the rapid burning of glucose by your muscles during and after a workout, may also help to reduce high blood pressure and to lower cholesterol. More noticeably, if you fall, you're more likely to simply get up again, maybe without even a bruise, and frankly, strength training will make you less likely to fall in the first place.

To be honest, outside of stimulating muscle and bone growth, promoting insulin sensitivity, and helping you maintain (not obtain) a healthy weight, every other benefit of strength training is in dispute. For every study stating the positive effect of strength training on hypertension or heart disease, there's another that says there's no correlation. Read enough studies and scholarly articles, and all things will be both true and false at the same time. Frankly, this quantum superposition effect is a bit discouraging, and it makes every *Time* magazine article and CNN Health segment immediately suspect. But that's fine; the known benefits of strength training are more than enough.

Well, known to me now. Back in June of 2017, I knew none of this. I just wanted to be jacked. At the time, I had it in my head that I didn't like gyms. They always seemed a little like a scam to me, designed more to extract money from your bank account in perpetuity than to get you in shape. Then there would be all the slender people, sweating and judging and standing way too close. Rather than join a gym, therefore, I reached out to some guy on Craigslist and picked up a pretty nice weight bench, an Olympic-sized bar with 100 pounds in plates, plus two dumbbells that could hold 35 pounds each. I brought it all home and assembled it in the attic.

For the next few weeks I made the lifting motions I imagined bodybuilders made, but periodic trips to the full-length mirror in my bedroom made it clear that my imagination was lacking. I was managing to add mass to my upper arms, but otherwise I looked ridiculous. There was no muscle definition in my chest, shoulders, or anywhere else, and I was still pudgy around the middle. I obviously had something wrong. To find out what, I dove into Google.

Days later, I tried to synthesize all that I had learned. For one, the look I was going for is called "the classic V," a narrow waist and broad shoulders. You won't be surprised to learn that to get that shape you need to do specific exercises targeting specific muscles, and there are lots of these exercises. Indeed, there exist thousands upon thousands of different ways to twist, stretch, push, pull, bend, fold, and mutilate the human body in order to make it stronger, and there are a thousand more ways of arranging these exercises to achieve some specific goal. You can, and people do, dedicate a lifetime to fully understanding it.

At this point in our story, the beginning of July 2017, I managed to cobble together a new workout plan, which in hindsight was pretty good. The routine consisted of different exercises for different muscles, which all by itself put it light years ahead of my previous non-routine. It had push exercises (like the bench press) and pull exercises (like the barbell curl), it had reps and sets[48], it even had separate upper and lower body days.

I was still lacking two important pieces of information, however. One, most bodybuilding exercises require proper form, that is, they need to be executed in a very certain manner to be safe and effective. Unfortunately, proper form is rarely intuitive, which meant I was doing almost all of these exercises incorrectly. And two, I was missing the basic notions of muscle overload and progression. And it's progressive overload that causes hypertrophy, the opposite of atrophy, that is, muscle growth.

Protein Synthesis

Remember that steak and peach meal we had back in Chapter 2? It's time to catch up with the meaty half of that steak, the protein half. Last we saw of it, the protein had been broken down into a bunch of amino acids and passed through the wall of the small intestines. We'll pick up the story from there.

The amino acids of all types — there are 21 of them, remember — come trickling through the intestinal wall. From there they take a direct ride to the liver along a special portal vein that unlike most other veins in the body, does not lead directly back to the heart. At the liver, the amino acids branch out into the capillary system surrounding it, and are taken inside.

Recall that amino acids are the building blocks of proteins, and that proteins are *everywhere*. Enzymes are made from proteins. Chemical messengers, i.e., hormones, such as insulin, are made from proteins; so too are the receptors on the cells listening for those messengers. Chemical transporters, such as hemoglobin, are made from proteins. Antibodies are proteins. Skin, bones, hair, and every organ in your body are made from proteins. Muscle cells, in particular, are dense with proteins. It's estimated that a single human cell contains between one and three billion proteins! Most of the amino acids in that steak will be turned

48 Do an exercise once, say, curl a dumbbell, and that's a rep, as in repetition. Collect a bunch of reps together, say, eight or ten, then pause, and that's a set. And, yes, I had to specifically look up even that.

back into proteins, that is, turned into you, but amino acids can also be used for energy, or, if in the liver, can be turned into glucose.

The liver is an incredible machine, responsible for over 500 different metabolic processes, from storing glucose, to making bile, to removing toxins from the blood. We'll revisit the liver when we discuss metabolizing monosaccharides and lipids. As pertains to amino acids, however, one function of the liver is gluconeogenesis, the creation of glucose. If the body is low on glucose, such as when you're on a low-carb diet, the liver can synthesize it from certain amino acids (or other sources). The liver will also synthesize some non-essential amino acids from the amino acids you just ate. Finally, the liver can grab a bunch of amino acids and use them to create various proteins found in our blood, such as albumin[49] and various clotting factors. About 50% of the protein you eat is ultimately processed by the liver. Any amino acids not pressed into service by the liver, which will include all of the essential amino acids, the kind the body cannot synthesize, are dribbled back into the bloodstream.

These free amino acids will enter into cells throughout the body, where they will be put to work, mostly performing protein anabolism. That's a fancy way of saying "making proteins." Anabolism is the process of building up complex organic molecules from simpler ones. Such a process is said to be anabolic.

There are roughly 37 trillion cells in the human body[50], which is just an eye-popping number, and each of those cells is carrying around at least one copy of the source code for making a new human. Not just any human, of course, it's the code for making a new you. This code can be found in the nucleus of the cell in the form of deoxyribonucleic acid, DNA. The term "nucleic acid" is again one of those legacy names. A somewhat more accurate, if less mellifluous name for DNA would be "deoxyribose polynucleotide," since the constituent (monomer) parts of DNA are nucleotides. A nucleotide is a complex molecule that has a nitrogen base hooked on to a five-carbon sugar hooked on to one or more phosphate groups. The sugar component, deoxyribose, is a monosaccharide derived from another sugar, ribose, by the loss of an oxygen atom — it is 'deoxygenated.'

When strung together, the sugar-phosphate component of each nucleotide forms the backbone of our polynucleotide — okay, nucleic acid. It's the nitrogenous bases (aka nucleobases aka bases) that are the information bearing

[49] Albumin plays several roles in the body. As we'll see later, it's very useful for carrying other molecules around in the bloodstream. It also plays a critical role in maintaining the *osmotic pressure* of your blood. As blood travels through narrow capillaries, hydrostatic pressure forces nutrient rich plasma into the interstitial space. Albumin, then, maintains the pressure gradient needed to pull that water back in. You're likely more familiar with albumin in its egg white form.

[50] https://www.tandfonline.com/doi/full/10.3109/03014460.2013.807878

component of DNA. As you likely know, in DNA there are just four bases: cytosine, guanine, adenine, and thymine; C, G, A, T. Somewhat simplistically, a contiguous subset of these bases describes how to assemble a particular type of protein. That protein describing stretch of bases is called a gene, although not all genes describe proteins.

To describe a protein, a gene needs a "start" marker (a promoter), a "stop" marker (a terminator), and an ordered list of amino acid descriptors in between. Because there are 20 amino acids that get used for protein synthesis (the remaining human amino acid has another use), we'll need three consecutive nucleobases in the descriptor in order to uniquely specify each one. Since there are four different bases, DNA can be thought of as having a base-4 numbering system. One "digit" can have four distinct values (C, G, A, T). Two digits can have 4×4, or 16, values, which is not enough to uniquely specify all 20 amino acids. Thus, we need three digits, $4 \times 4 \times 4$, or 4^3, or 64 values, to represent them all. It's a bit on the overkill side, but the redundancy is useful.

But DNA is just the code. When a cell needs to manufacture a protein, and they pretty much always need to manufacture proteins, they call on other machinery to transcribe the code onto a new medium and carry that copy to a machine that knows how to read it and can assemble the new protein from the raw material of amino acids.

The first step is transcription, making a copy of the protein-building instructions. Here, an enzyme called RNA polymerase is put into action, scanning the length of the DNA for the promoter that says "instructions for protein X start here." Ignoring the double helix stuff and a few other complexities, once the enzyme finds the promoter, it copies the instructions (a particular sequence of C, G, A, and T nucleobases) onto another type of nucleic acid called RNA, ribonucleic acid. In particular, we call this nucleic acid mRNA, where the "m" stands for "messenger." We call it that because the mRNA has to leave the nucleus (by binding to proteins, of course) and take the "build this" message out into the cytosol[51] of the cell.

There's an interesting quirk to mRNA in that it doesn't use thymine in its copy of the DNA sequence. Instead it uses a similar but less energetically expensive nucleobase called uracil. Therefore, the four bases used to code for a protein in mRNA are C, G, A, and U. Recall that we need three bases to uniquely specify an amino acid. These nucleotide triplets are called "codons." Messenger RNA has a special "start" codon, AUG (i.e., adenine, uracil, guanine) that also codes for

[51] Cytosol is the aqueous substance that makes up most of a cell. Cytoplasm, on the other hand, is the cytosol plus everything that's suspended in it, such as the nucleus and mitochondria.

the amino acid methionine. In addition, it has three different "stop" codons, which do not code for any amino acid at all. All the other possible codons specify individual amino acids. Since we have many more combinations of C, G, A, and U than we need, some amino acids are coded for by several codons. For instance, the amino acids arginine and leucine are both encoded by six different nucleotide triplets.

Let's imagine a peptide that is just three amino acids long. Let's further imagine that this is long enough to be considered a protein, which it most definitely is not. We'll call this protein "lacein," after its discoverer (inventor) it. If the three amino acids that make up lacein were, oh, phenylalanine, glutamate, and leucine, then the DNA sequence that codes for this protein might be TTTTTAGAA, although other representations are possible. Once transcribed over to mRNA, the sequence that encodes lacein would begin with the start codon, AUG, followed by the slightly altered (Ts become Us) three-codon representation of our protein, UUUUUAGAA, followed by one of the stop codons, say, UAG. Putting it all together, the instructions for building a lacein protein are AUGUUUUUAGAAUAG (which, coincidentally, is also the sound a Nazi would make if you used your fast twitch muscles to punch him in the throat). We can make those lacein-building instructions slightly more legible with some hyphens: AUG-UUU-UUA-GAA-UAG. Which in English is: Start with the amino acid methionine, attach a phenylalanine then a glutamate then a leucine, then stop. Simplicity itself.

Inside a typical human cell are a bunch — and by "bunch" I mean millions — of little protein factories called ribosomes. The ribosomes will snatch up a snippet of mRNA from the cytoplasm, maybe one that describes the actual protein myosin, one of the principal proteins found in muscle cells, and threads that into its reader. Nearby, there happens to be another kind of nucleic acid called tRNA, "t" for "transfer." Transfer RNA is named as such because it transfers amino acids from the cytoplasm into the ribosome. There are 61 different types of tRNA, one for each possible amino acid-coding triplet on the mRNA strand (64 - 3 stop codons).

Even though we passed over the double helix nature of DNA, at this point we need to mention that the four nucleotide bases are complementary and able to form weak hydrogen bonds with each other. That is, they attract each other. In DNA, guanine always attracts and pairs with cytosine and adenine pairs with thymine. Once transcribed to mRNA, of course, adenine pairs with thymine's replacement, uracil, instead. Knowing this, we can now note that a tRNA molecule is effectively dual sided. On one side it has an anticodon string that pairs with one of the codons on the mRNA, and on the other side it has a binding site specifically for the amino acid that the codon describes. So, if a stretch of mRNA has the codon for leucine, GAA, then it will attract a tRNA molecule that has the anticodon, CUU, allowing them to chemically bond together.

Here's where we bring it all home. The mRNA is threaded into the ribosome where the first amino acid, methionine, is put into place as the start of a protein. The next codon, say, alanine's GCU, goes by the ribosome's read-head. This matches up with a tRNA that is already holding on to an alanine. This alanine-specific tRNA is drawn into the ribosome. The previous amino acid, methionine, is plucked from its tRNA cradle, allowing its carboxyl group to covalently bond with the amino group of the newly arrived alanine, releasing some water and forming a very short peptide, a dipeptide. Then the next codon is read and the process repeats: the corresponding tRNA is slotted into place, the growing peptide chain is transferred to the newly arrived amino acid, and the previous tRNA molecule — now missing its amino acid — gets booted out of the ribosome to pick up another. This process continues until the stop codon is reached and the protein is complete. For the hormone insulin, it repeats 51 times, but for certain muscle-related proteins, such as titin, the process will repeat as many as 33,000 times!

When this polypeptide, this protein, has been fully translated from its mRNA description to an actual chain of amino acids, it typically undergoes some post-translational modification where bits (like the introductory methionine) might get snipped off or other bits (like phosphates, lipids, or carbohydrates) might get attached or the whole thing may get cut in two (like with insulin). Ultimately, though, the newly minted protein goes SPROING! as all of its constituent molecules attract and repel each other in interesting ways. This causes the protein to fold and twist and turn until it resembles nothing more than a strand of Christmas tree lights after Daddy's had a few. The protein is then either put to work inside the cell or, if it's, say, a hormone, passed into the bloodstream to go and do its job. In the case of myosin, it grafts itself into place inside a muscle cell, where it helps with muscle contraction.

That's what happens to the vast majority of the protein you eat. It gets broken down into amino acids, distributed throughout your body, and turned right back into protein again. Any unused amino acids, and there won't be many if you eat right and exercise regularly, eventually find themselves back at the liver, where they are either put to use doing liver-y things, broken down and pissed out[52], or

[52] One of the byproducts of decomposition is ammonia, a toxin which is eliminated from the body primarily as urine. This makes some people assume that a protein-rich diet can damage the kidneys. Fortunately, this is not the case if you already have healthy kidneys. If not, you may need to dial back on the protein.

converted into glucose — very little of which enters the bloodstream immediately, but is instead stored as glycogen or fat. Importantly, the body doesn't store raw amino acids like it does carbohydrates and fats. It can synthesize some, and if necessary, break down other proteins for their raw materials, but in essence, amino acids need to be continuously replaced via your diet.

As a general rule, non-athletes need to consume around 0.4 or 0.5 grams of protein per pound of ideal (not actual) body weight, daily. Based on the standard, if simplistic, BMI model, my ideal body weight is around 170 pounds. A year ago, I would have needed between 68 and 85 grams of protein a day to meet my body's maintenance needs. An athlete needs even more protein, between 0.5 to 0.8 grams per pound per day. If you're looking to build muscle, you will need still more; approximately 1.0 to 1.5 grams of protein per pound per day. Since I actually am looking to build muscle, I need to be consuming between 170-255 grams of protein per day[53]! For comparison, Americans on average consume roughly 100 grams of protein a day, and that ribeye has about 90 grams all by itself. And all that extra protein that athletes and bodybuilders require, it's all being turned into muscle by your ribosomal protein factories.

Muscles are, of course, made up of muscle cells, also known as muscle fibers. As cells go, muscle fibers are on the large side, three centimeters long on average. There are muscle cells in your thigh that are as much as 30 centimeters (nearly 12 inches) long! Muscle fibers are rather thick, too; only female egg cells are wider. But what makes a muscle cell a muscle cell is not its length or its width, it's all the protein contained within it. A muscle cell's cytoplasm is simply choked with the stuff. There are many different kinds of proteins to be found in muscle tissue, but two predominate: myosin and actin. These long, fibrous proteins interleave with each other to form the basic contractile unit of skeletal muscle, the sarcomere. Sarcomeres, in turn, stitch themselves together end to end to make filaments called myofibrils. It continues on fractally from here. A bunch of myofibrils get packed into a muscle cell (fiber) along with all the other cellular machinery: nucleuses (plural!), mitochondria, ribosomes, etc. These muscle fibers are themselves wrapped together into bundles called fascicles. Finally, a bunch of fascicles are wrapped together into what you would recognize as muscle. Threaded around the fascicles are the fuel-delivering capillaries and the signal-delivering nerve cells. If you've ever seen a cross section of a transatlantic fiber optic cable, you have a pretty good mental image of a muscle.

---

[53] I can't be bothered tracking my actual consumption, but I do bias my diet toward protein. If I had to estimate, I'm probably consuming around 200 grams of protein a day.

Strength Training

A complete explanation of the increase in muscle size that bodybuilders experience is surprisingly elusive. But it is generally accepted that the act of adding muscle mass is primarily due to myofibrillar hypertrophy. That is, adding additional sarcomeres (made of the proteins myosin, actin, etc.) to the myofibrils of existing muscle fibers. But not evenly. Slow twitch, type I fibers, the muscle tissue tuned for endurance, do not experience much hypertrophic growth at all. It is largely the fast twitch, type II muscles, the ones tuned for power, that respond to strength training. Bear in mind that muscle mass does not increase due to the addition of new muscle cells, but only by making the existing cells larger.

Whether you're male or female, in order to add muscle mass, if that is indeed your goal, you need to stress the muscle. More than stress it, actually, you need to really traumatize it. That is, induce muscle exhaustion by asking more of your muscles than they can handle. Not too much more — that's pointless and dangerous, as we'll see — just a little more. When your muscles get stressed this way, they get damaged, and several hours later, your body will start repairing that damage.

Or maybe not. While the notion of muscle damage and repair is repeated endlessly as the cause of muscular hypertrophy, like so many other things concerning muscle growth, the received wisdom may not actually be the case. It seems that just taxing your muscles, and not necessarily damaging them, is enough to send a growth signal to your cells so that they get larger and stronger in anticipation of being taxed again.

There is a hormonal component to muscle growth as well. Various anabolic hormones are responsible for signaling the need for cellular repair, promoting protein synthesis, adding additional blood capillaries, and delivering glucose and oxygen to fuel the whole process. Through a complicated signaling pathway, strength training stimulates the release of the steroid testosterone. Testosterone has a lot of roles to play in the human body, but when it enters a muscle cell, it binds with a so-called androgen receptor. Androgen receptors are proteins that, once activated, dive into the cellular nucleus and tell it to start making muscle proteins. That is, testosterone is a regulator of gene expression. It's why muscle cells make muscle proteins and not, say, kidney proteins.

In the end, given the right diet and level of exercise, your muscles will become larger. Slowly. It takes repeated bouts of strenuous exercise for growth to become visibly apparent. As with the loss of water weight when dieting, you'll have an initially positive experience called "newbie gains." Newbie gains are increases in muscle size due to better utilizing the muscle fibers you already have, like inflating a flat tire. These are real gains, and will likely have a very positive impact on your self-esteem, don't think of them as somehow fake. However, after the rapid initial growth, the process slows down dramatically, and you will need to spend months working out to continue to add muscle mass.

There is a great deal of misinformation and disagreement when it comes to the process of building muscle. How much should you lift? How many sets? How many reps? Do you need sets at all? What should you eat? When should you eat it? What exercises should you do? What exercises should you stop doing? Every conceivable question is asked and answered in every conceivable way. It's to the point where it seems that all this information is simply conjured out of thin air just to deliver ad impressions. That said, professional bodybuilders and all those guys at the gym *are* adding muscle, so some of this received wisdom must be at least directionally correct.

Assuming you're only seeking to add enough muscle to look good in a tee-shirt, and not to enter any bodybuilding competitions, here's the absolute minimum of what is currently considered good strength training practices. First, you have to stay at it. It will be at least six months, and more likely twelve, before there will be any appreciable muscle gain. (Remember, though, that simply staying healthy and preventing muscle loss requires strength training for life.) Second, you can't rest too long between workout days; you'll want to hit those muscles one to three times a week. Third, there needs to be sufficient weight load to trigger a response; approximately 75% of what you would be able to lift just once. That is, if you could curl a 100-pound barbell once, then you should be doing reps of curls with 75-pound barbells, less if you're adding reps. Put another way, the amount of weight to use is the amount needed to reach exhaustion around the last rep of each set. And that implies the need for reps, 6-10, and sets, 2-4, and resting 2-4 minutes in between. Finally, you'll need to target all of the major muscle groups in your body with both push and pull exercises.

Once you've dialed in to a good routine, there will come a day where you can do one more rep. You've gotten stronger. This is where we introduce the notion of progressive overload, the real driver of hypertrophy. When you are finally strong enough to finish all reps in all sets, then you increase the weight, usually by adding 2.5-5 pounds. You need to force your muscles to adapt to ever-increasing workloads. As they do, they will get bigger and stronger. And you will be a little bit closer to the classic V. Increasing the weight is the most common way to induce overload, but there are many others; you can increase or slow down the speed of each rep, you can do more reps in a set or do more sets, you can reduce the time between sets, you can practice your form or simply concentrate harder on your routine. The key is to progressively increase the level of effort.

It is vitally important that you do the exercises properly, thus reaping the benefits and avoiding injury. I suffered my one and only injury in August of 2017, an injury I'm still dealing with now, twelve months later, because I didn't know what I was doing. This is especially sucky when you're 53 years old and exercising for the first time. Not because age keeps you from healing or anything like that, but because you're not broken. People my age who have been exercising regularly for years, have a lifetime of injuries to deal with. I didn't. The world was my oyster, until I twisted my knee.

The details aren't important, but following some bad advice found on the Internet, I added too much weight to the bar (mistake) when doing a barbell squat. A barbell squat entails putting a weighted barbell across your shoulders, squatting down, and standing back up. Because I was doing this in my attic, and thus lacking the proper equipment (mistake), I had to get that barbell from the floor onto my back (mistake). Which was fine for me when the bar weighed 65 pounds, but at 105 pounds, I had to jerk the weight to get it over my head (big mistake), and that made my knee do something knees shouldn't do. To drive home exactly how wrong one can be, given my level of fitness and experience at the time, that naked 45-pound bar should not have had 60 pounds added to it. Even the 20 I was using previously was wrong. I should have been doing those squats with an empty bar!

There are several takeaways from all that. One, don't believe everything you read on the web; try and corroborate what looks legitimate with other sources. If that's not possible, use common sense. When that fails you, and it will, ask yourself, "Can this hurt me?" If it seems the answer is yes, then don't do it. This advice is useful far outside the realm of fitness.

Two, don't cheat. Perform all exercises through their full range of motion. Don't yank on the weights; that is, don't rely on momentum to complete just one more rep. If you take on more weight than you're actually ready for, you do nothing to promote muscle growth and everything to promote injury. Lower the weight and perform the exercise as intended. Myself, I only stopped cheating in January of 2018. Six months after embarking on a weight-lifting regimen, I reset everything and concentrated on completing the exercises with proper form and at the proper weight. The results were immediate. I hurt less, I looked better, and, honestly, my workouts were more enjoyable.

And three, if you are doing any significant strength training at all, and that training requires free weights, then don't work out in your freaking attic! You won't have enough weights, you'll spend half your life moving plates from bar to bar, and, as we've seen, you'll risk injury when you don't have the right equipment for doing something as simple as a squat.

Feeling Stronger Every Day

In the middle of August, two weeks after hitting my (then) target weight of 180, four weeks since quitting smoking, six weeks past my first fledgling attempts at bodybuilding, and three months after embarking on all of this, I was in South Lake Tahoe, California. I've always had a fondness for California, ever since I was nine or ten and first saw *The Parent Trap* on TV. When the movie finally gets to Brian Keith playing the rancher dad in Monterey, I was in love. I was in love with his house, his land, his job, and I have to say, in a totally heterosexual way, his barrel-chested, squared-jawed physicality[54].

We'd rented a nice place right in the Tahoe Keys, which, isn't anywhere near Monterey, but still nice. Beautiful, in fact. California's drought had recently ended in spectacular fashion, and the lake was full. The weather was perfect (indeed, back in drought conditions, but that wasn't official yet), the birds were singing, and God was in his heaven. It was wonderful.

The one thing California didn't have was my weight bench. We arrived on a Wednesday, and the next day was supposed to be upper-body day. What was I going to do? I couldn't not lift for an entire week, I'd lose all my gainz. Bodyweight exercises weren't going to cut it. What to do? What to do? (None of that is true, mind you, but I didn't know that at the time.)

Denise informed me that a gym nearby took drop-ins. I informed Denise that we'd been married for 19 years, and she should know me better by now. Whereupon she let me know that the guy she was looking at was not the same guy she married and maybe this changeling version of me would like going to the gym.

Her logic was unassailable. On Thursday afternoon, I went to the gym. Ninety minutes later, she and I were having the following conversation.

"It's unbelievable," I said. "They have row after row of dumbbells, and they go heavier than 35 pounds. Much heavier."

"I know, honey," Denise replied.

"They have machines that I can use to do those exercises I've read about, but can't do in the attic."

"I'm sure they do."

---

[54] Brian Keith also developed smoking-related lung cancer, but killed himself, possibly for unrelated reasons, before it could kill him.

"There's this cage-like thing that has a barbell in it. You can use the cage to get the barbell on your shoulders so you can do squats."

"It's called a squat rack."

"They even have something you can use to do pull-ups."

"I've been to a gym before, you know."

"And best of all, there was hardly anyone there!"

We'll revisit California in the Chapter 7. For now, it's enough to note that I spent two more afternoons in that gym, watching the routines other people did, marveling (as I still do) at the crazy amount of weight some people can lift, and trying to figure out how the machines worked. It was clear, I'd need to join a gym after all. And I'd need to find a better workout routine.

Upon my return, I found a gym, the Vanguard Key Club, coincidentally located in the same building as Fuel, the boot camp people. It's a small, no frills gym biased towards strength training, although there are aerobic machines, as well. But there's no pool, no juice bar, no sauna, and, best of all, no crowds, not even staff. To access the facility, you simply swipe a little key card that they send you.

While waiting for that key to arrive, I dove back into Google with renewed vigor. I needed a workout routine designed primarily to add muscle. Now that I was somewhat more versed in weight-lifting arcana, it didn't take long. One gentleman, known only by his first name, Jay, stood out in a very crowded field. Maybe it's due to his conversational (okay, repetitive) writing style, maybe it's because of his commonsensical approach to diet and exercise, or maybe it's because everything is so neatly packaged, but the information on his site[55] looked reasonable to me and, it seems, with others in the know.

I'm not knowledgeable enough to say with certainty that Jay's workout routines are well designed or if the reasoning behind them is correct, but I can say that they are at least as effective and correct as anything else out there, and likely a great deal more so. I've been on *The Muscle Building Workout Routine*[56] for almost twelve months now, and it's working as advertised. However, even though everything you need to get going is either on Jay's website or in his book, you may benefit from the following additional hard-won advice.

You are a beginner! Follow the beginner routine that Jay makes available[57]. Don't worry, you'll still get your gains. To build up your strength and to get your

[55] https://www.aworkoutroutine.com/
[56] https://www.aworkoutroutine.com/the-muscle-building-workout-routine/
[57] https://www.aworkoutroutine.com/the-beginner-weight-training-workout-routine/

body used to this new track, take the time necessary to learn each exercise's proper form (YouTube helps a lot here, standard caveats apply). I wish I had. Maybe then I wouldn't have a bum knee. Maybe I wouldn't have had to do a complete reset of my workout. Maybe I wouldn't have wasted so much time not getting the results I'm getting right now.

The intermediate level protocol that I erroneously selected actually consists of four routines: two different upper body routines (Monday and Thursday) and two different lower body routines (Tuesday and Friday). These provide some variety and the opportunity to hit different muscle groups (upper: chest, back, shoulders, biceps, and triceps; and lower: quads, hamstrings, calves, glutes, and abs) with different exercises.

I walked into the gym for the first time on August 28th. I was pleased to see only a small handful of people there. Earlier, I had copied the routine into a note-taking app on my phone. That day was a Monday, so I was going to work through the "Upper Body A" routine. The first exercise was three sets of 6-8 bench presses. There are three benches at Vanguard, and at least one is free at any one time. I chose one, added 70 pounds to the empty bar (for a total of 115 pounds — the bar weighing 45), lay down, and did six quick reps. It was a little too easy, but I figured I should stick with it and adjust the weight next week. I waited a couple of minutes, lay down again, and did six more. Then I did it again.

Not too shabby, I thought. But it was. I did no warm up sets, which would have prepared my muscles, tendons, and mind for what was about to come. I did the exercises too quickly and didn't use the full range of motion. That bar should have touched my chest on the way down, and I'm pretty sure it didn't. I know it didn't later, as I started to add load. The point is, focus on safety, form, and learning at first, don't just Leroy Jenkins into this.

Anyway, the rest of that evening, and frankly, every other workout since then, went by in much the same way. I'd work my way through that day's routine, and at the end of each set, I would jot down the weight load in a simple notation specifying whether I should increase it (+) or not (~) the next week. After a while it would look like this:

```
Incline Dumbbell Press, 3 sets of 8: 40+ 45+ 50~ 50~ 50+ 55~
```

That notation allows you to see progress at a glance, but for some reason I could always keep pull-ups in my head. Three sets of six pull-ups are the first exercise on Thursdays, that is, "Upper Body B" day. By pull-ups, I mean palms out, full extension, chin-to-the-bar pull-ups. None of that namby-pamby half-ass shit. On August 31, the first day I went through this routine, I approached the pull-up station, grabbed the bar with an overhand grip, and did exactly zero pull-ups. Zero! I subbed in the lateral pull downs as recommended.

Every week I tried again, slowly making progress. One pull-up, two pull-ups, etc. On December 30, four months after my first attempt, I was able to do six pull-ups in my first set, followed two minutes later by four and four again. On January 18, my set count was 6-6-4. However, it wasn't until February 21 that I was able to complete three sets of six pull-ups. But I get ahead of myself.

As expected, I got my newbie gains, and felt pretty good about myself. After that, progress was slow but enjoyable. On Mondays, Tuesdays, Thursdays, and Fridays I would get home from work, change into my workout gear, drive to the gym, and go through the day's routine. (This is on top of the morning cardio I describe in the next chapter, by the way.) As I got more familiar with the gym, the routines, and my body, I made some small changes. For example, switching Dumbbell Flys (which hurt my shoulders) for Cable Crossovers (which don't). But for the most part, I followed the routine closely, and still do a year later.

I won't bore you with a blow-by-blow of what four days a week in the gym is actually like, just a few observations.

- A lot of people in this world are just ridiculously strong. There are guys in my gym who routinely deadlift 400 pounds or do ten reps of 150 as their bench press *warm-up* set. It's crazy. But it doesn't upset me to know that I'm the weakest guy in the gym. I'm an old man who's been at this for only a year, and even so I'm putting on a small arms gun show. I'll never be as strong as these guys, but it's okay; I'll be strong enough.
- Gyms aren't only for the young, as I imagined they would be. At my gym there's at least one octogenarian, and quite a few quinquagenarians (look it up) like me. Most gym rats are pretty fit, not unexpectedly, though there's a fair number of people who are trying to slim down. At Vanguard, I'd say men outnumber women by about 2 to 1, probably due to the emphasis on free weights.
- Among people who are knowledgeable in this space, there's a general preference for free weights over machines, because in order to maintain balance and stability, free weights make you call on many more muscles than the ones primarily being targeted, while machines tend to isolate particular muscles. That said, my current regimen uses both equally.
- As I mentioned, if your workout routine includes chest flys, consider replacing them with cable flys (aka crossovers). Not only do you get to pretend you're Hercules or Samson, but they're safer and more

effective. Also consider swapping out the skull crushers for tricep dips on an unassisted dip machine. Besides really bringing on the swole, when the dips get too easy, you can wrap 30 pounds of chain around your waist to make yourself heavier, and that's pretty badass.

- The intermediate routine rightly tacks abdominal exercises onto the end of leg day. For no good reason, after about six months or so, I decided to slot ab exercises *in between* sets on *every* day. Which, if nothing else, makes leg day shorter.
- If you have to travel occasionally for work, don't worry too much about messing up your routine. If it's just one day, you can make it up on one of your off days, or not, no biggie. If you're on the road for several days, then the hotel fitness center is typically good enough; a few dumbbells and that contraption with all the pulleys and cables will keep you in tune. If you have to travel for a full week or more, then Google a local gym; most allow drop-ins for a fee. But honestly, don't sweat it.

◆ ◆ ◆

All was good through October, until, as already related, I decided to try and lose those ten extra pounds. As we now know all too well, in order to build muscle, your body needs something to build it out of, namely, an excess of protein. But when you're dieting and amino acids are scarce, your body will scavenge the raw materials it needs from the protein-rich tissue that constitutes nearly half of your makeup. Now, you can try and discourage this muscle catabolism — the opposite of anabolism, think "cat" as in "catastrophe" — by continuing to exercise hard and convincing your body to look elsewhere for amino acids, like, I don't know, the spleen or something. Spleens don't do anything important, do they? But that only minimizes the loss, it doesn't stop it. In short, when you are restricting calories, there will be some amount of muscle loss.

My failed November diet didn't really restrict calories, however, it just made consuming them more annoying. That is, I'd cut back calories in the morning, but have an excess at night. Or I would starve myself on Monday only to gorge myself on Tuesday. My calorie consumption yo-yoed up and down like that for a month before I quit. My performance at the gym suffered commensurately; I lost zero pounds and gained zero muscle.

Between Thanksgiving and Christmas, once I switched back to a consistent calorie load, everything began to normalize. It appears that the mental and physical toll of almost-but-not-quite dieting is huge. But with my meals and caloric consumption back to normal, my spirits were high and my body was rebounding. Unfortunately, having not yet put two and two together, and as discussed in Chapter 3, I started dieting again shortly after Christmas. When, a few weeks later, that diet also failed, again leaving me physically unchanged, I didn't just throw in the towel, I also threw in the bathrobe and the shower curtain.

I had burnt out. Primarily, I think, because of the maddening winter of 2017-2018, with the chilling cold and the endless dark and the (previously unmentioned) three consecutive weeks of mind numbing, soul deadening, spirit crushing all-day offsites at work[58]. The cold and the dark and the work commitments had destroyed my cardio regimen. On top of that, add the drain of my disastrous attempts to lose those last ten pounds, and I was depleted of my final reserves of energy. I took a break from weight lifting.

Now, it's perfectly okay, even beneficial, to take a week away from the iron every few months, but I did it at the worst possible time: while I was restricting calories — again. Recall from Chapter 3 that immediately following my second attempt at losing those ten pounds, I started a third, ketogenic, attempt. But by not working out at the same time that I had cut calories to about 1,200 a day, I was basically telling my body to start tearing down muscle to get at the amino acids stored there. In other words, there was muscle loss — mirror-visible muscle loss. But, hey, I lost ten pounds, if only briefly.

Hitting the reset button wasn't the end of the world. A little muscle loss while I mentally regrouped was worth it. I even managed to take a brief vacation someplace warm in February. Soon enough, I was back at the gym in earnest. In order to ease myself back into things, I knocked the load down for each exercise by about 25%. It took a few weeks, but I got back to where I was before Christmas, recovering enough to finally notch three sets of six pull-ups. When, in May, I started my fourth, final, and successful attempt at dropping ten pounds, I even managed to avoid the worst of the muscle degradation by staying on point at the gym and supplementing my meals with drinks of milk and protein powder.

It's late July now, a little more than a year since I started working out in the attic. Can I deadlift 400 pounds? Am I jacked like Michael B. Jordan? The answer to both questions is no, of course not. Not even close. Perhaps a better question, given my goals, is how do I look?

---

[58] I mean, honestly, exactly how long can people discuss the role of a "Scrum Master" in an agile organization? Seemingly forever, it seems. Here, let me help you out. If you are ever asked what role a Scrum Master should play on a software development team, answer "none." There is no place for a Scrum Master, because there is no place for Scrum. Scrum is not an effective software development methodology. It doesn't work. It has never worked. In fact, Scrum makes software worse! There should not be Scrum Masters in an organization in the same way that there should not be witch doctors in a hospital. If senior management at Mass. General ever asked what role shamans should play in the hospital, the only possible answer is: "None. No role. Witch doctors have no place in a hospital." Christ, why are we still having this conversation?

To be honest, the results of a year at the gym are less visible than I had hoped. There are guys at my gym with chests, shoulders, and arms that beggar belief. Even my more modestly sized gym-mates are clearly in a different league than I am. That said, a few of these strongmen are also carrying around some excess weight[59], while I, at 6'1" and 170 pounds, am lean. Along with being lean, in my case, anyway, comes not being broad-shouldered. It's a tradeoff. Even so, my shoulders are broader than my waist, so there's that.

Diet is not the only reason that I won't be cast as Josh Brolin's body double any time soon. Age has something to do with it, too. As a man in his mid-fifties, I'll never be able to add muscle the way a twenty-something can, or even a fifty-something who's been lifting for 30 years[60]. Moreover, it has only been a year since I started strength training. I don't know what I'll look like another year hence. Regardless, when it comes to a body's response to bodybuilding, there is one thing more important than diet, age, practice, skill, grit, determination, or plain ol' orneriness, and that is genetics. It is extremely likely that I, my father's son, am simply genetically lean. Sinewy, even.

But lean is not synonymous with scrawny. In fact, for many athletes, acquiring lean muscle is the goal. I suspect that if someone were to see me in a nicely tailored dress shirt, and felt obliged to comment on my appearance, they might say, "He's well turned out." And if they were to see me in a snug fitting tee, they might be inclined to comment, "He's surprisingly fit for a man his age." But if they were to see me on the beach or at the pool, they might — just might — have occasion to remark, "Dude, that old guy's pretty cut." I can live with that.

---

[59] Just as the normal consequence of losing weight is losing muscle, the normal consequence of adding muscle is gaining weight.

[60] Before you say it, yes, Josh Brolin is only four years younger than I am, but he has a Hollywood-supplied personal trainer. I suppose the CGI doesn't hurt, either.

Chapter 6: Burn

In order for the human body to do anything at all, it needs energy. That is, it needs chemicals to react rather violently together. It gets that reaction primarily by oxidizing glucose and producing, among other things, a chemical called adenosine triphosphate, or ATP. We'll discuss later exactly how ATP equates to energy, but first we have to make some. To do that, we need to catch up with our peach and the fatty half of that steak.

Metabolizing Monosaccharides

The peach we ate way back in Chapter 2 is slowly being broken down into glucose and fructose, and these monosaccharides are trickling through the intestinal wall on their way to the liver. I used words like "slowly" and "trickle" to reinforce the idea that the simple sugars found in whole fruit, being all mixed up with the fiber, are qualitatively different from the sugars found in a soft drink, which are anything but slow. Eat a peach, and roughly 10 grams of sugar will make its way into your bloodstream over the course of a few hours (a few more grams will feed your gut biome or get pooped out). Drink a Coke, however, and 39 g of sugar will start hitting your bloodstream in perfect harmony — and 20 minutes.

While it varies from individual to individual and from morning to night, if you haven't eaten in a few hours, the typical amount of glucose in the bloodstream is about four grams, or in the range of 70-90 mg/dL, and your body really, really wants to keep it that way. Accordingly, there's a number of metabolic systems that all coordinate to keep glucose levels in that range, storing it when there's too much and releasing it again when there's too little, even hoarding it at times, and, if necessary, manufacturing glucose from other sources. When you drink that Coke, your body needs to quickly remove pretty much all of the sugar from your bloodstream, leading to insulin spikes and overly aggressive lipogenesis (turning sugar into fat) by the liver. Dealing with the peach, on the other hand, is a much more leisurely affair.

Put another way, a peach has a glycemic index of 28. Recall that the glycemic index (GI) is a measure of how *fast* a particular food causes blood sugar to rise, on a scale of 0-100, with 100 being pure glucose. In contrast to a peach, a 12-ounce Coke has a GI of approximately 59. Interestingly, a healthy, delicious watermelon has a GI of 76. Thus, it seems the glycemic index isn't telling the whole story. The reason for this seeming disparity is that some foods, like watermelon, have very little sugar, but release it very quickly. To account for the amount of sugar *and* how quickly it's released, there's a more useful value called the glycemic load (GL). On the GL scale, 0-10 is low, 11-19 is medium, and 20 and above is high. A Coke has a GL of twenty, while a slice of watermelon has a GL of four[61]. For the most part, the foods with a high GL are all the foods you suspect already: soft drinks, bagels, candy, pasta, and so on, while the low end of the scale is dominated by fruits and vegetables. Keep in mind, that size matters, one grape has a lower GL than ten grapes and one ounce of soda has a lower GL than twelve.

The monosaccharides from the peach pass through the intestinal wall and into the hepatic portal vein, taking the same path to the liver as the amino acids took in the last chapter. From there, they travel to the heart, then out into the circulatory system. When the pancreas senses the rise in glucose[62], but not fructose, it secretes a hormone known as insulin, via the wonderfully named islets of Langerhans — scientifically known as pancreatic beta cells. Insulin's primary role is to help stabilize the amount of sugar in the blood, trying to keep it around the four-gram mark. It does this by telling all the other cells to open up, sugar's coming.

What happens next depends on the cell, but of immediate interest is the liver, which starts joining the glucose molecules together into a substance called glycogen, and stashes that glycogen away for when glucose levels start to drop again. Glycogen is thus the short-term storage form of glucose, the analogue of starch in a plant. The liver will continue to make glycogen so long as the flow of sugar continues, that is, so long as that peach is being digested. There is an upper bound, however, to how much glycogen it can store: about 100 grams. If there's still glucose in the blood after that, the liver will switch over to turning glucose into fat, a process known as lipogenesis. This fat is bound up with some proteins

[61] For the record, a peach has a GL of five.

[62] Free fatty acids and certain dietary amino acids also trigger insulin secretion, but to a much smaller degree.

and cholesterol into little balls of lipids called very-low-density lipoproteins, VLDL, and shipped off to the adipose tissue for long-term storage. In other words, you've just gotten a tiny bit fatter. If more sugar arrives before these reserves can be depleted, you'll get fatter still. Moreover, the liver can make fat faster than it can secrete it, meaning that if the sugar keeps flowing, the liver itself will get fatter. Hello, non-alcoholic fatty liver disease, a relatively new chronic condition brought on by the rise of obesity across the globe. When the aliens finally come "to serve man," I hope they have a taste for foie gras.

The liver doesn't capture all of the glucose passing through it, only about 20% at any one time. The rest continues on into the bloodstream. As it flows through capillary beds and around hungry cells, it can be taken into those cells and converted into energy. But glucose cannot pass through the cellular membrane unassisted; a glucose transporter protein is needed to allow the glucose in. It's insulin's job to cause these proteins to cluster at the cell membrane and provide that pathway[63]. If your cells are not responding to insulin, the glucose can't easily enter the cell, and it continues to float around your body while your cells starve. We call this type 2 diabetes.

If the glucose-consuming cell happens to be a muscle cell, then it is inclined to take in a little more glucose than needed to meet current energy demands. The muscles will turn this excess glucose into glycogen, just like the liver did, and hold on to it for later use. This glycogen is especially useful for when the muscles need energy in a hurry. There's about 500 grams of glycogen stored in the muscles of a healthy adult, a little more than a pound. Together with the glycogen stored in the liver (100 grams), you're looking at about 2,400 calories (600 grams of carbohydrates × 4 calories per gram) worth of stored glucose, enough to keep you moving for about a day if not replenished.

Besides triggering glycogen production in the liver and muscles, insulin also signals your fat cells to start taking in glucose and converting it directly into fat. That is, your fat cells are also capable of lipogenesis, just like your liver, and now you're even fatter. In addition, insulin triggers the adipose tissue to release a hormone of its own, called leptin. That's right, your fat cells, technically known as adipocytes, are part of your endocrine system. Like insulin, leptin has a number of jobs, but job number one is to bind to receptors on the hypothalamus, a region of the brain which, among other things, coordinates the autonomic nervous system regulating body temperature, thirst, hunger, and other homeostatic systems. The arriving leptin signals the hypothalamus that more fat is being made, which has the consequence of making you feel less hungry and of raising your

---

[63] The release of adrenaline during exercise has the same effect of stimulating these glucose transport proteins.

base metabolic rate. That is, leptin induces a feeling of satiety, or to be somewhat more accurate, leptin, a slow acting hormone, signals the brain that the danger of starvation has lessened so please get off the couch. It's also why eating slowly generally makes you eat less: the leptin has a chance to reach the hypothalamus and send the "I'm full" signal before you can devour your meal.

There's a related hormone called ghrelin involved as well. Ghrelin is produced primarily by the stomach in response to it being empty (the tap is metaphorically closed again when the stomach expands). Like leptin, ghrelin binds with receptors on the hypothalamus. In fact, it tickles exactly the same cells as leptin does, as those cells have receptors for both hormones. However, when ghrelin activates these appetite-controlling cells, they make you feel hungry, and it does so faster than the "you're not starving" message that leptin sends. Incidentally, ghrelin also stimulates a dopaminergic reward response in the brain, making the desire for food hard to resist.

In general, this appetite modulating system works well. But the caloric density and palatability of the typical 21st century Western diet can overwhelm these governors. As you get heavier and heavier, the leptin signal gets more and more attenuated, leading you to eat even more and move even less. The end result is leptin resistance and the inability of the body to properly regulate food intake and energy expenditure. Worse, since your new normal is "fat and slow," people who are leptin resistant but still manage to lose weight have a much harder time keeping that weight off.

Such is the current understanding, anyway. Leptin was only discovered in 1994, and a great deal is not yet known about its mechanism of action or what other regulatory roles it may play. In particular, it is not clear what actually causes one to become leptin resistant, only that it correlates with obesity. Of even more interest is that there may be no such thing as leptin resistance at all. We may have it all backwards, and it's possible that raised leptin levels do not signal "you're full" as much as *reduced* levels signal "you're starving." That is, the signal is only meaningful when leptin levels are low, and one needs to find food and conserve energy. Health science is tricky.

All of this metabolic processing is happening continuously in the hours it takes to digest a meal. Eventually, though, some six to eight hours later, digestion will be complete, and all of the accessible nutrients will have filtered through the intestinal wall. Now, imagine this happens as you're driving from Boston to Chicago. Since Toledo's already behind you, you decide to ignore the hunger pangs, and instead thumb the cruise control up to 80 MPH — you'll eat when you get to The Weiner's Circle. Go for extended periods without eating (or exercise hard enough) and your blood sugar levels will begin to fall, which, from your body's point of view, is not good at all. Too little glucose, hypoglycemia, is bad. Deadly bad. Therefore, the body essentially reverses course and starts undoing what it was doing when glucose was abundant. First, the pancreas shuts off the insulin tap and releases another hormone, glucagon, instead. Glucagon, released from the pancreatic alpha cells, is essentially the anti-insulin, just as ghrelin is the anti-leptin. It tells the liver to stop making glycogen and start breaking it down, turning the glycogen back into glucose and releasing it into the bloodstream. If glycogen supplies are low, it tells the liver to start *making* glucose out of fatty acids and amino acids. It should be noted that the glycogen stored in muscles is not used to normalize blood sugar levels; that glycogen is kept solely as a backup energy supply for the muscles themselves.

Glucagon also signals both the liver and adipose tissue to stop making fat and start releasing the stored fatty acids back into the bloodstream, instead. We'll see shortly how cells can burn fatty acids instead of glucose, thus saving the now-precious glucose for use in the brain and red blood cells, neither of which can take advantage of fatty acids as a fuel source. This also explains why you stop feeling hungry after a while of ignoring the original hunger signal: your liver is essentially feeding your cells from stored sugar and fat.

Along with the glucose in the peach, we also ingested a small amount of fructose. Fructose, glucose's very close cousin, is metabolized very differently. Most significantly, it is not used directly as fuel. Whereas glucose is metabolized by cells throughout the body, fructose is metabolized almost entirely in the liver, where it suffers various fates. Somewhere between 30 to 50 percent is enzymatically converted into glucose and released into the bloodstream. Approximately 25 percent more is converted into lactate and similarly released into the circulatory system, where it too can also be used as cellular fuel. Another 15-18% is turned into glycogen and stored locally in the liver. Finally, a small fraction of the fructose consumed will be converted into triglycerides, that is, fat. Besides not acting as a fuel source, fructose also doesn't trigger an insulin response — why bother, the cells can't burn it — therefore, it doesn't trigger a leptin response, and thus, no "I'm full" signal is generated. At least not directly. Since most fructose is quickly converted into glucose and glycogen (stored glucose), fructose is indirectly doing all those things, just slower.

It's worth discussing why the body stores glucose as both glycogen and fat. Why not store only glycogen? After all, glycogen is simpler for the body to make and unmake. The reason is that fat is much more energy dense, with nine calories per gram, as opposed to just four calories per gram of glycogen. If all excess calories were stored as glycogen, you would be carrying around more than twice as much excess weight as you are now. That is, glycogen is essentially your quick release, short-term energy store, and fat is your more compact, long-term energy store. Furthermore, glycogen is hydrophilic, attracting water. Fat, as we'll see next, is hydrophobic. This means you can store a ton more fat without the additional water weight.

Metabolizing Triglycerides

Along with the amino acids and sugars, fatty acids (the other half of our steak) are seeping through the intestinal wall as well. But fatty acids are metabolized completely differently from anything else. First, right there on the far side of the intestinal wall, the fatty acids are remade into triglycerides. Because lipids (fats) are hydrophobic, the triglycerides and a little cholesterol are then bound up into a ball and coated with a layer of phospholipids. The phospholipids are very similar to the triglycerides they enclose, except that one of the fatty acid tails attached to the glycerol backbone has been replaced with a phosphate group. The tails face in toward the triglycerides, but the hydrophilic phosphate head faces out, allowing our little ball of fat to be dispersed into the blood. This sphere, awesomely known as a chylomicron, is also studded with a small number of proteins that later will assist certain enzymes in metabolizing it. In almost every respect, these chylomicrons are just like their smaller but more densely packed cousins VLDL (mentioned earlier), LDL, and HDL. In fact, the other name for chylomicrons is ultra-low-density lipoproteins, ULDL, but that's boring.

Now capable of navigating the aqueous environment of the bloodstream, but being too large to pass through the capillary network, the chylomicrons are shunted into the lymphatic system, where they find their way up to a transfer point near the neck and from there into the bloodstream via the much, much wider subclavian vein. Eventually, the chylomicrons will run through a capillary bed which conveniently holds the enzymes needed to shred their outer shells and break the glycerol bonds of the triglycerides. This releases the fatty acids for intake into nearby cells, likely the muscles, heart, liver, or adipose tissue. If it's the latter two, the fatty acids are turned right back into triglycerides and stored for a rainy day.

The stage is set. The curtain's rising. At this point we have our respiratory system supplying the oxygen, the digestive system supplying the fuel, the circulatory (aka cardiovascular) and lymphatic systems carrying the oxygen and fuel around, and the endocrine system orchestrating everything via the controlled release of hormones. All of this finely tuned machinery, this symphony of wonder, is at your beck and call, allowing you to perform physical and intellectual feats of extraordinary grace and power. To become the Earth's — nay, the galaxy's — apex predator. Or, you know, catch up on *The Bachelor*. Whatever. Doesn't matter. It's time to make some energy. Enter cellular respiration.

Cellular Energy

As mentioned briefly, the cell's source of power is a molecule called adenosine triphosphate, or ATP[64]. Although adenosine has a remarkable number of physiological roles to play, it's really the triphosphate part that generates all the excitement. A phosphate is a phosphorus atom surrounded by four negatively charged oxygen atoms. If two phosphates bind to each other (a diphosphate), it's a rather stable arrangement, like two young lovers who only recently met. What this couple really doesn't need right now is for one of their ex-partners to join them for any length of time. That arrangement, like a triphosphate, would be very unstable, indeed. Sure, it can happen if you try hard enough, but it's not going to end well. Yet that's what's taking place in your cells, over and over and over again, as you convert the sugar in that peach into useful energy.

Already in your cells is a fair amount of adenosine *di*phosphate (ADP), whose two phosphate molecules are happily holding hands and enjoying long walks on the beach. To make ATP, as cells do continuously, they attach a free-floating phosphate ion (Pi), our metaphorical ex-girlfriend, to the ADP in a process called phosphorylation, which itself takes energy. This rather unstable (but pretending otherwise) group of phosphates then awaits an enzyme (in this analogy, I suppose this is the best friend) to come along and remove the phosphate, i.e. hydrolyze it, and SNAP! The last phosphate loses its shit, shouting something like, "How could you! You said you would love me forever. I hope you and the slut are happy!" and storms off, very energetically.

This, of course, leaves you right back where you started, with an ADP molecule and (once she calms down) a phosphate ion floating around in the cytoplasm. No worries, in this Sartresque version of hell for our star-crossed lovers, your cell simply (or not so simply, as we'll see) sticks the ADP and Pi back together, making ATP again, which will get hydrolyzed for its energy once more, then stuck back together again, hydrolyzed, stuck back together, and so on, and so on *ad infinitum*. The trick is to use less energy phosphorylating the ADP than the resulting ATP generates when it breaks back down into ADP + Pi. And for that, we're going to need some glucose, some oxygen, and a somewhat better understanding of enzymes.

---

[64] ATP is a nucleotide, by the way, just like the monomers of DNA and RNA.

Enzymes Again

Enzymes, you'll recall, accelerate chemical reactions. Reactions that may have taken hours, days, or even years to occur in the absence of an enzyme can be catalyzed in microseconds when the appropriate enzyme is present. Enzymes are proteins, often multiple proteins working together. When these proteins were created from the peptide bonds of amino acids (as described in the previous chapter) they folded into a very specific shape. One of the properties of that shape is the creation of an "active site." An active site is an area on the enzyme that is perfectly matched to the shape of the molecular substrate that the enzyme can alter, much like a lock is shaped to fit a specific key. But it's not just the shape that makes an active site active. Recall that amino acids have side chains that make one kind of amino acid distinct from the others. These side chains are especially relevant at the active site, where that particular arrangement of atoms aids in bonding with just the right substrate, similar to how the anticodons on tRNA attract and bond to the codons on mRNA.

One kind of enzyme is capable of catalyzing one kind of reaction with one kind of substrate to create one kind of product. For example, the carbonic anhydrase enzyme we encountered in Chapter 4 joins the CO_2 substrate it finds in your blood with H_2O to make the product carbonic acid, H_2CO_3. Consequently, there are several thousand different kinds of enzymes in the human body, millions of them in any one cell, and trillions of enzymes altogether. All of them floating around in the cytoplasm, waiting for their particular substrate to drift by.

Frequently enzymes need other raw materials, called cofactors, in order to complete the reaction. Cofactors are typically either a metal ion, e.g. magnesium or zinc, or an inorganic chemical compound, a coenzyme, generally derived from vitamins, such as niacin or folic acid. One cofactor found in many reactions is a molecule called NAD^+ (nicotinamide adenine dinucleotide[65]), a derivative of vitamin B_3. NAD^+ is an oxidizing agent, that is, it accepts electrons from other molecules. We'll have more to say about oxidation shortly. For now, just know that NAD^+ exists to aid in the many chemical reactions that require the removal of electrons. Another common cofactor when additional energy is required is ATP itself, the very molecule we're hoping to make.

Even though enzymatic reactions are fast, they are not continuous; instead they're regulated by the cellular environment. Particular enzymes may become more or less active depending on changes in temperature, acidity, and the

[65] As you can see, nucleotides are important macromolecules.

presence or absence of various molecules. For instance, the process that creates ATP is inhibited by the very presence of ATP, that is, it self-regulates so that we don't end up creating more than we need. Finally, it frequently takes multiple reactions to get from the substrate we start with to the product we really want. These reactions, called metabolic pathways, happen sequentially, with the product of one step forming the substrate of the next. As with ATP, it is typical for the final product of a pathway to also be an inhibitor of that pathway's first step. One such metabolic pathway is known as glycolysis, and that's the one that turns sugar into energy.

Glycolysis

When a cell needs energy — and some cells, like muscle fibers and neurons, need a lot of energy — it expresses a protein receptor that binds to the insulin coursing through your veins. This sets up a channel for glucose to come pouring into the cell. In the cell's cytoplasm, ten distinct enzymes, in a process called glycolysis, and using the energy provided by two ATP and two NAD^+, will catalyze ten distinct reactions to convert one six-carbon glucose (hexose) molecule into two three-carbon (triose) molecules called pyruvate. It also generates four ATP molecules (for a net yield of two) and two electron donors called NADH. Got all that? No? Let's break it down.

Glycolysis is a ten-step metabolic pathway for converting sugar into ATP, a cell's very short term energy store, as well as a few other byproducts. At each step, except the first, the product of the previous step becomes the substrate of the next. I'm going to go into a fair bit of detail here so that you can better understand how enzymes and metabolic pathways work. That understanding will allow you to more readily imagine what happens in the many pathways I've already glossed over and those still to come, and to have a better sense of what it means when you hear such things as "the liver turns sugar into fat."

In the first stage of glycolysis, an enzyme called hexokinase binds onto the glucose, grabs some ATP (acting as a cofactor in this case) that's floating around nearby, and plucks that third phosphate off of it. The enzyme then attaches the phosphate to the glucose molecule, thus phosphorylating it and releasing the energy stored in the ATP[66]. This turns the glucose into the intermediary product glucose-6-phosphate (G6P), which, among other things, prevents the glucose from diffusing back out of the cell. This first step is also a regulatory step; the

[66] Chemical energy is released when bonds are *formed*, not broken, even though casual language suggests otherwise.

presence of G6P inhibits the production of more G6P until the current batch is used up.

Glucose-6-phosphate is the precursor of a number of metabolic pathways, including glycogenesis (the creation of glycogen) and the synthesis of the ribose skeletons of DNA and RNA. But we want energy, that is, ATP, so the next step in the glycolytic pathway is to turn the G6P into fructose (F6P). In the third step, another ATP is invested via the phosphofructokinase[67] enzyme, which further phosphorylates the F6P molecule. Those two phosphates are making the underlying fructose substrate increasingly unstable, and that will allow steps four and five to cut it into two three-carbon sugars.

When step five completes, we're done investing in the process of creating ATP and can start the payoff phase. All of steps six through ten repeat twice, once for each of the two trioses created in the investment phase. In step six, another enzyme will grab onto two cofactors: NAD^+ and a phosphate ion. It will use the NAD^+ to pluck two hydrogen atoms — including their electrons — from one of those sugars, transforming the NAD^+ into NADH. The energy from that electron transfer will then be used to phosphorylate the sugar substrate with the Pi. In step seven, yet another enzyme transfers the Pi from the now unstable sugar to a nearby ADP, resulting in our first unit of ATP (two, really, since the other triose has been similarly transformed). Steps eight and nine monkey with things some more in order to destabilize the sugar again, so that step ten, the last step, has enough energy to slam another Pi onto another ADP to create another molecule (two, really) of ATP.

There you have it. We invested two ATP to get four. In the process, we also turned one glucose molecule into two pyruvate molecules, and we transformed two molecules of electron-poor NAD^+ into two molecules of electron-rich NADH. Despite how long it took to explain this process, in reality it is exceedingly quick. If you need energy fast, the glycolytic pathway is what you turn to. From Chapter 5 we are already aware of a part of your body that can use a fast-responding energy source, your fast-twitch muscles. We'll return to this fact a little later, but for now take note of three things. One, no oxygen was used during the process of glycolysis, that is, it is anaerobic. Two, we have some byproducts that need to be dealt with, the pyruvate and the NADH. Finally, a NAD^+ has been used up, and we're going to need more of that if we want to break down glucose again.

Despite the 100% return on investment that we get from glycolysis, that's chump change, really. Sure, it's enough to get in a few quick punches, but it's not

[67] It's pronouncing this enzyme that makes biochemists so much fun at cocktail parties.

going to take you the distance. To go the full 15 rounds, you'll need to do a lot better than paying two ATP to get four. Especially since that's only about five percent of the total energy potential of glucose. Thankfully, there's a metabolic pathway that will allow us to take the byproducts of glycolysis — in other words, to take the initial investment of two ATP — and produce as many as 34 additional ATP molecules. For this pathway to kick in, oxygen is absolutely essential, although it's only used at the very end.

Pyruvate Oxidation and The Citric Acid Cycle

Among the many organelles making up the cytoplasm of a cell are the mitochondria. Some 1.8 billion years ago, mitochondria were very likely independent bacteria, until one entered a host cell and formed a symbiotic relationship with it. The host, previously powered only by glycolysis, got a supercharger, and the mitochondrion got food and protection. Consequently, mitochondria have their own DNA, ribosomes, RNA, and other protein-building machinery, and are capable of dividing (and joining) inside the larger cell.

It's in the mitochondria that the multiple phases of aerobic cellular respiration[68] take place. The number of mitochondria in a cell varies widely by cell type and even by how fit you are. Red blood cells, for instance, have no mitochondria in them at all, and rely solely on glycolysis for energy. Whereas a liver cell has between one and two thousand mitochondria. Better than counting mitochondria, which in many ways can't really be done, is counting how much of the cytoplasm they comprise. In the cardiac muscle of the heart, that number is around 30-40 percent. In fast twitch, type II muscles, that number is a paltry 1-2 percent. Conversely, the cytoplasm of slow twitch, type I, endurance muscle is made up of around 15-20 percent mitochondria — if you're fit. If instead, you spend your days in front of a computer and your nights in front of a TV, then mitochondria are taking up just five or six percent of your type I muscle cells' cytoplasm. Think about that.

If you consider glycolysis the first stage of cellular respiration, then the second stage is pyruvate oxidation. As you have likely guessed from the name, the substrate for this stage is, conveniently, the principal product of the previous stage, pyruvate (CH_3COCO_2H). When ATP is in high demand, a transport protein will carry the two pyruvate molecules generated during glycolysis into the mitochondria. There, a complex of three cooperating enzymes, known as

---

[68] The term 'respiration,' in this context, means the process of converting nutrients to energy. It has nothing to do with the physiological meaning of transporting O_2 and CO_2 between cells and the atmosphere.

pyruvate dehydrogenase, chops off an unwanted carbon dioxide molecule from the pyruvate and boots it out as waste, forming a two-carbon acetyl group (CH_3COH). Acetyl groups form the backbone of many important organic molecules. For instance, back in Chapter 4 we discussed the neurotransmitter acetylcholine, but you are likely more familiar with products such as acetaminophen or acetylsalicylic acid (aspirin). Acetyl group in hand, the enzyme next reaches for a cofactor derived from vitamin B_5 and known (boringly) as Coenzyme A, or simply CoA. Using the energy extracted from oxidizing two more NAD^+ cofactors into NADH, it attaches the CoA to the acetyl group to make acetyl-CoA. Remember that name, you'll be hearing it again.

Even though pyruvate oxidation is just one step, it's an important one, because it allows the resulting acetyl-CoA to participate in the next stage on this process, the citric acid cycle, also known as the Krebs cycle[69] after the guy that discovered it, Hans Krebs, in 1937. There's a lot going on in the citric acid cycle, and you'll be happy to know that I'm not going to explain it all. I will, however, need to highlight a few steps that will be important later.

As you have probably guessed, the citric acid cycle is an enzyme-mediated metabolic pathway. In the very first step of the cycle, the Coenzyme A that was attached to the acetyl group in the previous phase is plucked right back off again. The mediating enzyme, citrate synthase, then attaches a four-carbon molecule known as oxaloacetate (remember that name, too) to the two-carbon acetyl group to make a six-carbon citrate, from which the citric acid cycle gets the first half of its name. The second, "cycle," half of the name comes from the fact that nine enzymatic transformations later, the final product is — wait for it — oxaloacetate. This oxaloacetate can then be attached to another acetyl-CoA, and the cycle continues.

All along the way, the enzymes involved are not only binding to the products of previous steps, but to any number of cofactors as well. As with many reactions, a fair amount of NAD^+ has been transformed to NADH. In addition, the enzyme involved in step six, succinate dehydrogenase, makes use of another oxidizing agent, flavin adenine dinucleotide (FAD) — derived from riboflavin, i.e., vitamin B_2 — to accept electrons from step five's product, succinate, transforming the FAD into the electron-rich $FADH_2$ (You're following all this, right?). When all is said and done, the two two-carbon glucose-derived acetyl-CoA molecules have been catabolized into six NADH, two $FADH_2$, four CO_2 molecules[70], and two more ATPs (technically, the very closely related GTP, guanosine triphosphate).

[69] Also known as the tricarboxylic acid cycle.

[70] Released into the bloodstream as waste and ultimately exhaled, as described in Chapter 4.

And while you may think that those last two ATP molecules were worth all the trouble, they're really not what this process was after. While each ATP is a powerhouse of energy, they are exceedingly small. In fact, you need to snap 100 thousand billion billion different triphosphate bonds to release just one calorie of energy. In other words, we're going to need a *lot* of ATP. Thus, the real purpose of the citric acid cycle was to generate those electron-rich NADH and $FADH_2$ molecules, because these molecules are going to feed the last stage in this process, the electron transport chain, and that's going to pay off big time.

Oxidation and Reduction

Before we get there, though, I need to explain something else you were taught in high school, but have probably since forgotten. As promised, we need to talk about oxidation. Above, I referred to NAD^+ as an oxidizing agent, but what exactly does that mean? What is happening when FAD is "transformed" into $FADH_2$? Why does the citric acid cycle go to such lengths to produce these electron-rich metabolites?

The process of oxidation is the removal of electrons from one molecule, the electron donor, to some other molecule, the electron acceptor, with a corresponding release of energy. Because they can induce other substances to give up their electrons, another name for an electron acceptor is an oxidizing agent (although oxygen is not the sole oxidizing agent, it's just common); it oxidizes other chemical species. NAD^+ and FAD are electron acceptors — oxidizers. The substance that loses electrons, that is, the donor, is known as a reducing agent. Because, in our case, NAD^+ and FAD have gained electrons, they are said to be reduced[71] (transformed) into NADH and $FADH_2$. Since, you can't have reduction (gain) without a corresponding oxidation (loss), the two together are called "redox" reactions. Frankly, the nomenclature is all a bit wonky.

That's all well and good, but in organic chemistry, with its large and complex molecules, you can't realistically track individual electrons. Indeed, it's uncommon for electrons to be torn from one element and transferred to another. Instead, we use the following simplifying definitions, which, if nothing else, make the terms oxidation and reduction more sensical. Oxidation is when a carbon-based molecule gains oxygen or loses hydrogen. Reduction is the opposite, when an organic molecule loses oxygen or gains hydrogen.

---

[71] A confusing term given that the "reduced" molecule has "gained" electrons. It's left over from the 18th century when oxidation was not well understood.

This works because different elements have varying degrees of "electronegativity," which is the tendency for an element to pull electrons toward itself. Oxygen is highly electronegative — as are most of the elements in the upper right-hand corner of the periodic table — and hugs electrons closely. (Fluorine is particularly affectionate. It's the Lennie Small of the elements. In other words, do not fuck with fluorine.) Hydrogen, on the other hand, has low electronegativity, and does not love its electrons nearly as much as oxygen does. Now, imagine a balloon full of hydrogen gas, H_2, that is, a hydrogen atom covalently bonded to another hydrogen atom. In this case, the shared electrons spend as much time with one hydrogen as they do the other. The individual hydrogen atoms and the molecule as a whole have no charge. The same is true of O_2, the shared electrons spend equal time with each of the oxygen atoms, and the molecule remains neutral. But throw in a spark (as happened on the Hindenburg) and see what happens. The energy from the spark disassociates the two oxygen atoms and the two hydrogen atoms of any number of nearby H_2 and O_2 molecules. The electronegative oxygen atoms pull the hydrogen atoms towards themselves, and as the bonds reform we get a fair amount of water, a little hydrogen peroxide, and a rather dramatic explosion. Notice, though, that in the resulting H_2O, the shared electrons are not shared equally. In fact, the electrons that make up the hydrogen atoms are now spending most, but not all, of their time near the greedy oxygen nucleus. In a sense, the oxygen has gained two electrons and the hydrogens have each lost there's. While in actuality, the oxygen hasn't removed the hydrogen's electrons at all, it's just hogging them[72].

The point is, there has been an effective, if not actual, transfer of electrons from one chemical, now oxidized, to another, now reduced. This can be very energetic, as with fire, or less so, as with rust. The ATP generation process is ultimately going to rely on a fairly energetic transfer of electrons, i.e., redox reactions, to supply the energy needed to attach a phosphate ion to an ADP molecule.

The Electron Transport Chain

What we ultimately want is to oxidize, that is, remove electrons (hydrogen atoms) from the newly created NADH and $FADH_2$ molecules (generating NAD^+

[72] Since the H_2O molecule still has ten protons and ten electrons, it remains electrically neutral. However, it has a slight negative charge near the oxygen atom where the electrons tend to dawdle, and a slight positive charge near the hydrogen atoms. In other words, water molecules are polar. It's the attraction between the negative and positive ends of water molecules that causes them to attract each other, that is, to form hydrogen bonds, and make liquid water.

and FAD again, as well as a handful of protons), and capture the resulting energy. That means those negatively charged electrons will need to flow over to an acceptor, in this case, good old oxygen. This oxidation process would indeed be energetic. In fact, it would be so energetic (think fire, not rust) that the cell would not be able to capture it all to produce useful work. The electron transport chain (ETC), the final stage in aerobic ATP generation, exists to slow down the oxidation process so that the energy can be captured and transferred over to ATP production.

Picture a sausage. The kind you might see hanging in an authentic New York delicatessen. That's a mitochondrion. A membrane — the casing — surrounds densely packed cellular machinery, including the enzymes that propel the citric acid cycle; the DNA, RNA, and ribosomes that allow it to manufacture proteins; and, of course, a whole host of other proteins, cofactors, and metabolites. Like the mitochondrion itself, these guts get a cool name: the matrix, what in an ordinary cell we'd call cytoplasm. This picture is incomplete, however. To get a more accurate representation of a mitochondrion, you now need to picture that sausage being stuffed into the skin of a much smaller sausage, forcing the original sausage to fold and twist and squish in order to fit.

That is, a mitochondrion has two membranes. The outer membrane is much like that of the host cell; it is made up mostly of phospholipids. The small gap between the two membranes, the intermembrane space, is in turn much like the cytosol of a cell; largely water. The larger inner membrane that encloses the matrix (our original sausage) is somewhat different; it is absolutely studded with enzymes. It's these enzymes that will be put to work making ATP, but in a manner unlike anything we've seen so far.

The enzymes we've encountered up until now have mostly been floating around in an aqueous solution waiting for their matching substrates to wander by. The enzymes responsible for oxidative phosphorylation, however, are very different. For one thing, they come in cooperating clusters of five. Four of these are actually enzyme complexes, multiple enzymes working together. The fifth — well, you're just going to have to wait for that one. Another thing that makes these enzymes different is that instead of floating, they are embedded in the inner membrane, with one half sticking out into the matrix and the other half rising into the cytosol of the intermembrane space. These five enzymes make up the electron transport chain, which, via a mediated series of redox reactions, will take the energy in the NADH and $FADH_2$ molecules and use it to slam a phosphate ion onto an ADP molecule. The first four enzyme complexes, presumably named by the same guy that came up with the name "Coenzyme A," are known as

Complex I through Complex IV, though I suppose the Roman numerals add a bit of flair. The complexes are very much the same in function, so we'll start with Complex I.

The energetic NADH that were spun out of the citric acid cycle (and glycolysis) bind to the active site of Complex I and are oxidized back to NAD^+. Inside the complex are a series of coenzymes and iron-sulfur clusters that act as redox centers. The electrons from NADH are handed off to the first of these, then each successive redox center oxidizes its neighbor to its figurative left before reducing the neighbor to its right. As the electrons flow through the redox centers of Complex I, they set up an electric current. The energy from this current is used — and this next bit is important, so pay attention — to take four protons (H^+) from the matrix and propel them into the intermembrane space.

Having completed their exceedingly brief journey through Complex I, the electrons are transferred to a coenzyme carrier protein called, pleasingly, ubiquinone[73] (aka Coenzyme Q_{10}, aka Q), resulting in a product known as ubiquinol, or just QH_2. The QH_2 carries the electrons over to Complex III, where a similar series of redox reactions takes place, resulting in four more protons being booted into the intermembrane space.

Unlike Complexes I and III, Complex II is not a proton pump. It is a second entryway into the electron transport chain. Incidentally, Complex II has another name, succinate dehydrogenase. If that sounds familiar, it's because you just saw that name a couple of pages back. It is the selfsame enzyme found in step six of the citric acid cycle, the one that reduced FAD to $FADH_2$. That is, even as the citric acid cycle is turning acetyl-CoA into oxaloacetate, it is also pumping electrons straight into the electron transport chain. And when these electrons arrive, they too are passed through a number of redox centers, before also being transferred onto a Q electron carrier and shunted over to Complex III.

Once Complex III has harnessed the energy of the four electrons coursing through it, it donates them to a new electron carrier, cytochrome C, which transports them over to the final protein complex in the chain, Complex IV. As in Complexes I and III, the electrons flow through a number of redox agents, generating enough energy to translocate four more protons out of the matrix. But there's something different about Complex IV. Right smack dab in the middle of it is held one of the greediest oxidizers we know, molecular oxygen, O_2, and all four, now slightly less energetic, electrons go racing right at it. This causes the O_2 to very briefly dissociate into two oxygen atoms with a satisfying but overly

[73] Interestingly, statin drugs block the formation of this enzyme which is why one of the (many) side effects of taking statins is muscle fatigue.

generous ten electrons each. At the same time, however, four more protons are pulled from the matrix, and they happily fill their empty valence shells with the extra electrons the oxygens are carrying around, thus reducing the oxygen into two H_2O molecules, i.e., water. In fact, by the time all of the glucose fuel has been oxidized like this, six oxygen atoms will have been paired with 12 hydrogen ions and set adrift.

Think of this the next time you hear that your body uses oxygen to burn fuel. What's really happening is that high-energy electrons in your food have been sent through a bewildering series of redox reactions, surrendering a bit of energy each time. The electrons race through this transport chain because the redox centers and electron carriers at each step have a slightly higher affinity for electrons than the step before. They zip along until they reach the terminal oxidizer, O_2, which releases a great deal of energy as it accepts the electrons and carries them away in the form of water. This oxygen molecule, of course, will be immediately replaced by another that was taken into the cell as part of the respiratory process described in Chapter 4. Without the O_2 the whole pyruvate oxidation, citric acid cycle, and electron transport chain shuts down, and your body will rely solely on glycolysis for ATP production.

Speaking of which, where is the ATP? And what's with all that business of proton pumping? It's time we get to the payoff.

The fifth in our cluster of five enzymes is very different from the others. Since it synthesizes ATP, it is very appropriately called ATP synthase, and it too is embedded in the mitochondrial inner membrane. You can picture this enzyme as a small electric motor attached to a shaft attached to a sphere. You can picture it that way because *that's exactly what it is*! The motor part straddles the inner membrane. Attached to the motor is an axle, more accurately, a camshaft, that projects into the matrix. And attached to the camshaft is a collection of proteins arranged more or less like tangerine slices. The whole thing resembling an egg-beater more than anything else.

Recall that on the far side of the membrane there are now a lot more positively charged protons than a few moments earlier, considerably more than there are on the matrix side. Just as O_2 diffuses into oxygen-depleted blood in order to equalize the pressure, the difference in protons and charge on either side of the inner membrane forms an electrochemical concentration gradient, essentially a voltage, such that the protons in the intermembrane space are just aching to flow back to the proton-depleted matrix. And they do, via the only conduit available, by travelling through the ATP synthase enzyme like water through a waterwheel, and causing the little motor to spin — at nearly 7,000 RPM! The motor, in turn, spins the camshaft, and the "cam" that attaches to the spherical end — the tangerine end — of the enzyme causes it to deform, opening and closing the gaps between the components of the sphere. When a gap is opened, it exposes its active site and binds with an ADP and a phosphate ion cofactor, and when it closes again, it slams the two together rather forcefully, phosphorylating the ADP and making ATP.

And that's how we take the energy stored in glucose and convert it into an energy source the cell can use. QED.

Incidentally, this last step, the ATP synthase step, is so remarkable that creationists have latched on to it as proof of supernatural intelligent design. A view toward which you can be just a tad sympathetic when you realize that evolution solved the cellular energy problem with freakin' proton-powered protein motors. But replicators gonna replicate, and given the right environmental pressures, they'll come up with some ingenious solutions as they do. The oxygen-using electron transport chain found in mitochondria today evolved some two billion years ago. Even that is descended from similar molecules that evolved nearly two billion years earlier still, almost as soon as the earth was formed. Proton pumps, electrochemical gradients, and ATP-synthesizing enzymes are to be found powering nearly every living thing from muskrats to mushrooms.

It is quite the journey from the glycolysis of glucose and the synthesis of acetyl-CoA, through the citric acid cycle, and down the electron transport chain. But as promised, the payoff is huge. When this entire process completes, your cell has catabolized a single molecule of glucose into as many as 38 ATP molecules! That's the paltry four ATP we netted out of glycolysis and the citric acid cycle, plus a maximum of 34 from the electron transport chain. In practice, however, the electron transport chain produces closer to 30-32 ATP, as some energy is lost to inefficiencies and to cellular housekeeping, such as moving metabolites and waste in and out of the mitochondria.

These ATP molecules are then shunted out of the mitochondria and eventually hydrolyzed (via any one of a family of ATP-hydrolyzing enzymes) for their energy. Maybe to make some small part of a muscle move, or to help create a protein, or copy a DNA strand, or just plain think — your brain is a voracious consumer of ATP. It all requires energy. Once the ATP is dephosphorylated, the resulting ADP and free phosphate, if not needed elsewhere, are very quickly shunted right back into the mitochondria, where the whole process starts again.

Thirty or so ATP molecules per glucose molecule is all well and good, but what does that translate to in the large? Exactly how much ATP in total is being produced via all this machinery? After all, I just told you that you need about 100,000,000,000,000,000,000,000 (10^{23}) of them to make a single calorie, and an adult male is burning through some 2,500 calories a day, so it seems we'll need quite a bit.

There are different ways of looking at this number, but all of them are equally astounding. A single ATP synthase molecule is making three ATPs per revolution. At roughly 7,000 revolutions per minute, that's 350 ATP per second. But there are thousands of ETC clusters on a single mitochondrion (that's one reason the inner "sausage" is all squished up, to increase its surface area and thus hold more ETC clusters) and potentially thousands of mitochondria in a cell; add it all up and average it out, and a single cell — just one cell — *is making about 75 million ATP molecules per second.* It varies widely by cell type and many other factors, but at any one time, the average cell contains one billion ATP molecules, and it burns through them all in one to two minutes. But why stop at a single cell? As previously mentioned, there are about 37 trillion cells in your body, most of them producing ATP: adding them all up gets us to roughly two sextillion (2×10^{21}) ATP generated *per second,* or 240 septillion (2.4×10^{26}) ATP per day, give or take an order of magnitude or two. Put another way, the average human burns through their own body weight in ATP every day — athletes, considerably more so. Fortunately for us, we don't have to carry all this ATP around; it's recycled over and over again from the roughly 50 grams of ADP/ATP we do have in our bodies.

Remember when you were in high school and had to read *The Grapes of Wrath,* and then, 500 pages later, you learned there was a movie? Well, I hate to break it to you, but all of the above could have been summed up as follows:

$C_6H_{12}O_6 + 6O_2 \rightarrow 6CO_2 + 6H_2O$ + heat + 34-38 ATP

Which is a lot more succinct, but hardly as edifying as spelling out the whole process in English. By the way, if you reverse that equation, you have the process of photosynthesis: 6CO2 + 6H2O + light energy → C6H12O6 + 6O2.

Cellular Respiration Gone Wrong

There are many ways that the process of converting glucose into useful energy can fail. Arsenic, for instance, interferes with the functioning of the pyruvate dehydrogenase enzyme, keeping it from creating acetyl-CoA and thus shutting down the citric acid cycle and the electron transport chain. But assuming you're not an ageing bachelor looking to rent a room in Brooklyn, what are some of the more common means for cellular respiration to go sideways?

Free Radicals and Antioxidants

Do anything a few septillion times a day, and you're bound to make the occasional mistake. When cellular respiration goes awry, the result is a free radical[74], and free radicals are nasty. To understand what they are and why you generally don't want them zipping around in your cells, we're going to need a smidge more chemistry.

A proper understanding of free radicals (and the octet rule, for that matter) requires knowledge of atomic suborbitals, electron spin, the Pauli Exclusion Principle, and more, but I'm going to handwave over all of that and simply state that electrons like to pair with other electrons. I mean, *really* like it. Out in the world, it's rare to encounter unpaired electrons among the biologically interesting elements. Instead we find that electrons have either paired with a neighboring electron within the same atom or bonded with electrons in another atom, forming a molecule. Thus, we have an atmosphere full of O_2 and N_2, and not just O and N.

Sometimes, however, these bonds get broken in such a way that an electron loses its pair. An atom or molecule containing such a beast is called a "free radical." Because of the electron's desire to pair, free radicals are very unstable, or put another way, they are highly reactive. A free radical will do just about anything to pair up its lone electron, including ripping an electron — rather violently — from some other poor molecule. Of course, this simply results in the creation of another radical that will in turn rip an electron off one of its neighbors. This chain reaction will continue until an electron donor is found that can give up an electron and still remain stable. An atom, molecule, or enzyme that can lose an electron and not become a free radical itself is known as an "antioxidant," since it stops the oxidation process.

From this description you can see that oxidation reactions (losing electrons) and reduction reactions (adding electrons) are essentially making free radicals all the time. Since the electron transport chain is nothing more than an enzymatically controlled series of redox reactions, removing electrons from one molecule and adding them to another, it is constantly creating free radicals as intermediaries. Fortunately, the enzymes in charge of these parts of the chain quickly pair these intermediary radicals with electron donors before any harm can be done.

Every once in a while, though, things go south. For instance, an electron might exit the transport chain prematurely and get dragged straight to that big

[74] Cellular respiration is just one source of free radicals. There are many others, including sunlight, pollution, smoking (of course), alcohol (sigh), radiation, heavy metals (with an 's'), pesticides, and other industrial chemicals.

electronegative O_2 molecule waiting at the end of the line. This reaction breaks one of the two covalent bonds that join the two oxygen atoms together, leaving the O_2 intact, but now with two unpaired electrons. The newly arrived electron then pairs with one of these dangling electrons, leaving the other one hanging, and in the process creating the free radical known as superoxide.

Superoxide is very reactive. So reactive, in fact, that the body's immune system actively manufactures superoxide molecules and uses them to slice and dice invading pathogens. It's when superoxide is created by mistake during cellular respiration that you have to worry — but not too much. Complex life has been bathing in O_2 for quite some time[75], and we've picked up a trick or two. One of those tricks is called "superoxide dismutase," which is an antioxidant enzyme that sits around in your cells just waiting for superoxide to show up. When it does, this enzyme catalyzes a reaction that turns superoxide into O_2 and H_2O_2 (hydrogen peroxide). The H_2O_2 itself, while not a free radical, is still reactive and also needs to be rendered harmless. For this purpose, there's another enzyme, catalase, to turn hydrogen peroxide molecules[76] into water and O_2.

So we're all good, right? No harm, no foul? Well, mostly. The electron transport chain tends not to fail, and when it does, there are mechanisms to limit the damage. But the important word in that last sentence is "limit." We are discussing a biological process that — let me repeat this — happens septillions of times a day; it is not 100% perfect. Sometimes free radicals escape their antioxidant enzyme keepers, if only briefly, and start ripping electrons off of neighboring proteins, lipids, and even DNA, especially that nearby mitochondrial DNA. Consequently, this oxidative stress has been implicated to one degree or another in a large variety of pathologies, including a number of cancers, Alzheimer's disease, Parkinson's disease, male infertility, atherosclerosis, and many, many more. Indeed, the accumulation of these thousands of little cuts is thought to explain why we age.

Insulin Resistance

Another way for cellular respiration to go wrong is for it not to happen at all. That, of course, is the end result of type 2 diabetes. If the cells aren't responding to the insulin signal, then they aren't pulling glucose out of the bloodstream, causing hyperglycemia. To bring the serum glucose levels back in line, your

[75] Approximately 2.5 billion years.

[76] Note the plural. Catalase needs two H_2O_2 molecules to make two H_2O and one O_2. Fun fact #1: A single catalase enzyme can break down millions of H_2O_2 molecules each second. Fun fact #2: When you pour hydrogen peroxide on a cut, the bubbles you see are caused by the oxygen released by catalases catalyzing.

pancreas will make even more insulin, essentially trying to jolt the cells into submission. Which will work at first, but as with overstimulated dopamine receptors, it becomes less and less effective over time. With timely and proper treatment, it's possible to partially or fully reverse this state of affairs. Left untreated, however, not only can insulin resistance become permanent, but the overproduction of insulin might actually damage the insulin-producing beta cells. There's no coming back from that.

As the additional insulin fails to rid the bloodstream of excess glucose, it will fall increasingly on the kidneys to filter out the sugar. This is a job they've had all along, but now it's moving into overdrive. If the kidneys get overloaded, they will purge the excess glucose by dumping it into urine. This concentrated urine osmotically draws water in, making the diabetic want to pee, and leaving him thirsty and dehydrated. If the kidneys still can't keep up, then the unprocessed glucose remains in the blood. This causes the blood to thicken to the point of becoming syrupy in texture, and can lead to an increased likelihood of stroke, heart disease, hypertension, and diabetic coma. In addition, this syrupy blood cannot easily reach your extremities or pass through capillaries, which can cause diabetic blindness as neurons begin to die, as well as neuropathy in the hands and feet. Nerve damage in the extremities makes it more likely for a simple wound to go unnoticed, and the thickened blood makes it difficult for the body to deliver white blood cells to the wound, which together increase the likelihood of infection, gangrene, and ultimately amputation. All the while, as your bodily systems founder, your cells are simply screaming for fuel — "We need glucose NOW!" — which drives the diabetic to eat a muffin or a bag of Milanos, making matters worse. It goes on like this for a while, one long cascade of bodily failures, until, of course, one of them kills you.

Alternative Fuel Sources

Assuming all systems are functioning normally, the human body does a pretty fantastic job of turning sugar into energy. But what if there's no glucose at hand when the cells need it? Maybe you're on a low- carb diet or fasting during Ramadan or maybe it's just been a few hours since you ate. In that case, there are a variety of other fuel sources that your cells can burn.

Fat

Remember, when blood sugar goes down, so too do insulin levels, and glucagon levels start going up. This is your metabolism's big reverse switch. When the liver senses the hormone shift, it stops creating glycogen and starts tearing it down, sending the resulting glucose into the bloodstream. If it's your muscles that need glucose, they will also start devouring their glycogen stores. Over in the adipose tissue, your fat cells have similarly stopped creating triglycerides and have started to rip them apart, releasing them into the bloodstream. (If you're wondering how hydrophobic fat is able to travel through the aqueous bloodstream, it's by hitching a ride on the albumin proteins which we touched on in the previous chapter.) Recall that with less insulin around to open up the cells for sugar, less glucose is going to get consumed by them. But low insulin and high glucagon (and a few other hormones) are the signal for cells to start taking in fatty acids, which they can also use for fuel. In other words, when glucose is low, your cells switch over to catabolizing fat.

As with glucose, this process begins in the mitochondria. The first step in converting fat into energy is called beta oxidation. The analogue of pyruvate oxidation, it involves stripping two carbons at a time from the fatty acid's chain, then combining them with coenzyme A to make — dramatic pause — acetyl-CoA. And just like the glucose-derived acetyl-CoA, these fat-derived acetyl-CoA molecules get shunted into the citric acid cycle, and from there ATP generation continues as described above.

In the pyruvate oxidation process, there was only enough carbon to make two acetyl-CoA molecules. But fatty acids are little more than long chains of carbon atoms, which means they have a lot more fuel to feed into the citric acid cycle. It varies depending on the fat type (long chain/short chain, saturated/unsaturated, cis/trans), but a common fat, such as palmitic acid, found in your steak and having 16-carbon tails, can produce a whopping 129 ATP molecules[77]. As with glucose, this process of burning stored fat generates carbon dioxide and water as waste products — the *only* waste products.

Did you catch that? You just got skinnier. What was stored fat has been converted into CO_2 and H_2O. This is the answer to the question, "Where does the fat go when I lose weight?" Nearly all of it, 86%, is exhaled as CO_2. The rest, the water, may also be exhaled or it is turned into sweat or piss or put to use as a solvent. I think about this a lot when I'm exercising.

Ketones

It gets better. Fatty acids will indeed be efficiently metabolized by most cells. Not the energy-hungry brain, however; fatty acids are too big to cross the

[77] Again, after housekeeping, the actual number is closer to 106.

blood/brain barrier. And not red blood cells, since they have no mitochondria. And not, as it turns out, our hardworking liver. Right now, in this glucose-poor environment, not only is the liver busy shedding glycogen, it's also busy with a process called gluconeogenesis, which basically amounts to running the ATP process backward to make the glucose that you're not eating.

Since the liver is preoccupied coping with a drop in glucose levels, when the fatty acids in the bloodstream arrive there, they do not get turned into energy as expected. Instead, they *begin* to go through the beta oxidation process, spitting out a bunch of acetyl-CoA that would normally feed the citric acid cycle, but then the process gets interrupted. This is because oxaloacetate, the cofactor needed in the first step of the citric acid cycle, is also a critical cofactor in the metabolic pathway of gluconeogenesis. When we are hypoglycemic, manufacturing glucose takes priority, and catabolizing fatty acids is inhibited. With the citric acid cycle shut down, the liver needs another mechanism for turning fat into usable energy, ideally energy the brain can use, and it has one in ketogenesis.

Just as gluconeogenesis is the synthesis of glucose from various other substrates, ketogenesis is the endogenous manufacture of so-called ketone bodies; three distinct (and not terribly well named) molecules that each happen to contain a ketone group. A ketone group, in turn, is simply a carbon double bonded to an oxygen with two carbon-based side chains (rest groups). Lots of molecules contain ketone groups, and we've already encountered several, including pyruvate, fructose, and oxaloacetate. The three ketone bodies — beta-hydroxybutyrate, acetoacetate, and, to a lesser degree, acetone — are special because there's a metabolic pathway in the liver, and only in the liver, that can convert acetyl-CoA to beta-hydroxybutyrate or acetoacetate (acetone, the simplest ketone and the same stuff you find in nail polish remover, is just the breakdown product of these other two ketone bodies). These ketone bodies can be used as cellular fuel as easily as glucose or fatty acids, except, conveniently, by the liver, which is forced to release them into the bloodstream. Just like glucose, ketone bodies have easy access to cells, including neurons in the brain and central nervous system, and are readily taken in, converted back into acetyl-CoA, and thrown into the ATP-generating furnace.

As mentioned, the process of creating ketone bodies is called ketogenesis. The state of having ketones in your bloodstream is known as ketosis[78]. If you're on a low-carb diet, you will flip back and forth from being in ketosis and not

[78] Not to be confused with "ketoacidosis," a fatal condition associated with type 1 diabetes, where persistent low insulin and high glucagon levels trick the liver into generating ketone bodies (and glucose) by the truckload — far in excess of a normal state of ketosis — and thus dramatically lowering the pH level of the blood.

throughout the day. If you're on an extremely low carb/high fat diet, i.e., a ketogenic diet, you will primarily, if not solely, be in a state of ketosis. Which is to say, you are now burning fat, lots of fat. As near as anyone can tell, a long-lived or even permanent state of ketosis is perfectly safe. In fact, in many cases, it's desirable, especially for people with epilepsy. Burning ketones instead of glucose in the brain has therapeutic effects not just for convulsive diseases, but for other neurological disorders as well, including migraines, trauma, Alzheimer's and Parkinson's disease, sleep disorders, brain cancer, autism, ALS, and multiple sclerosis. The exact mechanisms for the efficacy of the ketogenic diet is not well understood, but the 40,000-foot version seems to be that the brain, having evolved when carbohydrates were not quite so plentiful, simply excels at burning ketones.

You do not need to suffer from any neurological disorders to reap rewards from the ketogenic diet. As with cardiovascular exercise, the keto diet is proving to be broadly therapeutic. These days it is primarily recognized for its proven ability to promote weight loss — dramatic weight loss — although, as with neurological disorders, it is not completely clear why the ketogenic diet is so effective at weight control. The leading explanation is that people on a ketogenic diet simply do not feel hungry between meals, which reduces anxiety and the tendency to cheat, and thus the likelihood of regaining lost weight. It's likely that appetite suppression is due to ghrelin production being suppressed when you are in a ketotic state[79]. Also, and as previously stated, when in ketosis, the body simply becomes a fat burning machine. Beyond weight loss, studies have also shown positive therapeutic results of a ketogenic diet on managing or preventing heart disease, despite (likely, because of) the significantly higher amount of fat consumed. To be fair, ketogenic diets may also be implicated in an increased likelihood of arrhythmia. If you plan on adopting a ketogenic diet for a few months or even years, just be sure to stay hydrated, get the recommended amount of vitamins and minerals — especially electrolytes, and get plenty of fiber.

One other thing the keto diet is good for is managing diabetes. Recall that prediabetes is 100% reversible, and all you need to reverse it is a simple, sensible diet and some light cardiovascular exercise, just as described in the book you're reading right now. Furthermore, not only can type 2 diabetes be managed, for some people in some cases, it too can be reversed. A management plan for type 2 diabetes will always consist of at least three things: medicine, whether that be insulin injections or a cocktail of other drugs; diet; and exercise. The more you bring your weight under control and get your body moving, the easier it will be to manage your diabetes.

---

[79] https://www.nature.com/articles/ejcn201390

But here is what's absolutely maddening about the diet advice frequently given to people with type 2 diabetes. Go visit the website for The American Diabetes Association and click on their "Making Healthy Food Choices" link[80]. What does it say? Well, next to all the pictures of berries and eggplants and grapes and broccoli, you'll find text to the effect of "healthy meal plans ... include fruits and vegetables [and] lean protein foods." Which to me, anyway, is completely insane. Diabetes is a disease that can be characterized as having too much sugar in your blood, and the diet the ADA is recommending consists of more sugar! Sure, whole grains and fruits and vegetables have a much better nutrient and fiber profile than pizza, but they're still carbs and carbs are sugar! My advice — fortunately backed by a handful of studies, though unfortunately not nearly enough — is, if you are diabetic or prediabetic, switch to a ketogenic diet. After all, what better way is there to reduce the amount of carbohydrates in your blood than to eat as few of them as possible? Please keep in mind, however, that I am not a doctor, scientist, nutritionist or in any way qualified to offer diabetes management advice. Nevertheless, you should totally do it.

Amino Acids, Alcohol, and Fermentation

Enough about diabetes. Let's keep going with this. What if there's no glucose, no glycogen, no fat, and no ketones at hand when a cell needs to make some ATP? Not a problem. Remember those amino acids that were stitched together into muscles and hormones and enzymes in the last chapter? Or the ones that are just floating around in the cell's cytoplasm with everything else, just waiting their turn? Yeah, well, there are metabolic pathways that, depending on which amino acid you start with, can metabolize them into any number of citric acid cycle intermediaries: pyruvate, acetyl-CoA, etc. If they have carbon in them, they're getting stripped down and burned up for energy. Not much energy, a little bit less than glucose, but energy nonetheless. Burning amino acids is not uncommon, about 10% of your daily cellular energy production comes from amino acids.

Okay, what if there's no glucose, no glycogen, no fat, no ketones, and no amino acids around? My advice? Use the last of your energy reserves to take a swig of Irish whiskey. At seven calories per gram, a shot of Jameson will deliver about 100 calories. That should be enough energy to reach the fridge.

Last one. What if there's no glucose, no glycogen, no fat, no ketones, no amino acids, and no alcohol around? In that case, my heartfelt condolences, you're dead. To be more precise, you're dead of starvation.

---

[80] http://www.diabetes.org/food-and-fitness/food/what-can-i-eat/making-healthy-food-choices/

Still, there is one more variation on this theme. What if you have the glucose, but you don't have enough oxygen around to burn it properly? In active people, this is a fairly common occurrence. Or what if you're lifting weights and leveraging those fast twitch, anaerobic muscles? In both these cases, the body relies solely on glycolysis and its meager yield of two ATP for energy. Recall, though, that glycolysis also yielded two pyruvate molecules and two NADH molecules, and oxidized two NAD^+ molecules in doing so. If we're going to run more glucose through the glycolytic pathway, we're going to need to clean up. Fortunately, your cells have a pathway for this too, and that is fermentation.

Fermentation in humans is fundamentally the same as the fermentation process carried out by the lactobacillus bacteria when making yogurt and pickles (good pickles, not the vinegar-sodden crap you get in the supermarket). It's also similar to the fermentation process carried out by yeast, a fungus, when making Jameson, but instead of turning pyruvate molecules into ethanol and CO_2, fermentation in your muscles turns the pyruvate into lactic acid. Lactic acid fermentation is, of course, an enzyme-driven process. In this case, the enzyme lactate dehydrogenase takes the two NADH molecules made at the end of glycolysis and transfers their electrons right back to the pyruvate. This turns the pyruvate into lactate (lactic acid) and the NADH back into NAD^+. It's that last step we needed, so that the NAD^+ can be used to oxidize more glucose. That's it. Not much energy — just the net two ATPs from glycolysis — but, hey, better than being eaten by a tiger.

Unfortunately, this process has the side effect of leaving a bunch of lactic acid behind, which until recently was presumed to account for sore muscles in weight lifters, but is now known to account only for the aching immediately following a workout. Any lingering soreness is simply the result of muscle trauma, like after a rousing game of Punch Buggy. It's possible for the lactate to be used directly as fuel, but most of it finds its way back to the liver where it forms the fundamental substrate used for gluconeogenesis. That is, it's turned back into glucose and released into the bloodstream, where it makes its way back to the muscles and is either stored as glycogen, if your muscles have calmed down, or sent through another round of glycolysis.

We have now enumerated all of the mechanisms available to turn a rib-eye and a peach into energy, even if that energy has to be stored temporarily as glycogen, fat, or even tissue. As with what goes on inside a single cell, all of these bodily processes are happening at once. Don't think of the human body as a factory, think of it as a city. One of those well-run Nordic cities where everybody bikes to work. At any one time, some cells are burning fat and some are burning sugar, metabolic processes are started and abandoned, hormone levels rise and fall, proteins are being constructed and torn down. In aggregate, what the body is doing is trying to stay in the same steady state — homeostasis. It wants blood sugar, acidity, temperature, gases, minerals, and so on to be kept at equilibrium, and it is constantly using all the resources it has at hand to make the micro adjustments needed to do just that.

Chapter 7: Run

Newburyport is an extraordinarily attractive city. That this is so was not preordained. Back in the early 1970s, Newburyport, like many of the industrial cities that dot the Massachusetts landscape, went through a near-death experience. During America's otherwise golden post-war age, when (white) kids across the land were going to soda shops, rebelling without a cause, and listening to that demon rock and roll, Newburyport languished. Two hundred years after its founding, the city (not town, mind you, city) was on the verge of collapse.

The European ancestors of all those white kids first settled in the area in 1635, just 15 years after the Mayflower made landfall. They called the area Newberry Plantation, later changing it to just Newbury. That someone else already lived there was of little concern, especially since the indigenous population had been decimated by years of internecine warfare as well as the viruses brought over by previous waves of European settlers, and were in no position to contest the land grab.

Over one hundred years later, in 1764, the densely populated region of what was otherwise farm country cleaved itself off from its surroundings, and Newburyport was born. Newburyport's location near the natural harbor formed by the estuary of the Merrimack River, turned it into a thriving shipbuilding and trading community. In fact, at the time of the American Revolution, Newburyport was every bit the metropolitan rival of Boston, 35 miles to the south.

As you may imagine, by virtue of its age and location, the city is steeped in history. It was here that the United States Coast Guard was founded in 1791. It was in the neighboring town of Newbury where Benedict Arnold mustered the troops for his ill-fated invasion of Quebec. From Newburyport's shipyards, a whole host of privateers were launched to harass British shipping during both the Revolution and the War of 1812. George Washington visited on several occasions, and there's an oversized statue of him just outside my front door to commemorate that fact. Indeed, the entire cast of *Hamilton* has visited at one time or another, as you would know if you read the plaque outside our gorgeous public library: "Here Were Entertained George Washington, Thomas Jefferson, John Quincy Adams, Marquis de Lafayette, Benedict Arnold, and Aaron Burr." The latter two, presumably, when they were still being invited to parties.

While there were many economic ups and downs, shipbuilding, silversmithing, tanning, and several other important industries, not to mention all that plundering, kept Newburyport ticking along for almost 200 years. Much of the wealth that was generated (or liberated) went into the construction of the stately Georgian and Victorian homes that line High Street to this day.

This boom and bust cycle continued until the 1920s, when Newburyport began a long slow cycle of bust and bust. The days of the wooden ship were over, Boston won the trade port war, the factories were closing, and the British were the good guys now, which put the kibosh on all that plundering. The death knell came when the Interstate Highway System and the age of the automobile arrived in the 1950s, and people's newfound mobility started to gut the remaining downtown businesses. By the late 1960s, with many of Newburyport's majestic homes preserved only by "benign neglect," and the buildings surrounding Market Square suffering far worse (contemporaries likened Newburyport to Dresden after the war), the city started down the road of razing the downtown area and putting in — of all things — a strip mall.

To be fair, one needed real vision to see the beauty of Newburyport in 1970. Thankfully, those visionaries existed, and working with local, state, and federal governments, managed to preserve and restore the downtown. This successful public investment sparked additional private investment in restoring those great, neglected homes. Which in turn led to a larger tax base (the euphemistic way of saying "gentrification"), which led to additional investment, and so on. Such that today you can stroll down our brick-paved streets, past our beautiful Federalist and Greek Revival buildings, around our Olmsted-designed parks, and along our riverside boardwalk.

All of this is to say that Newburyport is a lovely place to take a walk.

Do the Walk of Life

By the end of June 2017, both boot camp and the 10,000 Step Challenge were done, but I wasn't. At this point, it would have been trivially easy to pat myself on the back and say, "Job well done," then go back to my slothful ways. But, surprisingly, such an act never occurred to me. I wasn't about to give up the near constant sense of delight that exercise gave me. I just needed to come up with something else to do.

It wasn't just the euphoria that kept me moving. I knew then, as everyone does, that exercise was "good for you," and a new desire to get healthier and skinnier was definitely part of the reason for maintaining an exercise regimen. I even somehow knew that I needed at least 30 minutes a day of aerobic exercise — the kind that gets your heart rate up — for the exercise to benefit me. I'm not exactly sure how I knew this. It's the kind of thing you just pick up. In much the same way that I know there's an entertainer named Cardi B, but not a damn thing else about — presumably — her.

It was easy enough to decide on what exercise to do; I would simply keep up my morning walks. I'd have to find something else too, of course, something a little more up-tempo, but for now, let's focus on the walks. They were wonderful. Most mornings were bright and cool, and I would get up early to hit the actual bricks. Sometimes with Denise, more often alone, which is fine; I don't mind being alone. I tried a few new routes. One is the Clipper City Rail Trail, a revitalized section of abandoned railroad track that leads from the not-at-all-abandoned commuter rail station all the way up to the river. There it splits, with the right fork allowing access to the pedestrian crossing on the Route 1 bridge. From there it crosses the Merrimack and joins up with the similarly revitalized Old Eastern Marsh Trail in Salisbury. If you walk from my house to the Rail Trail entrance on Low St., then the length of this trail system and back again, you will have covered just under five miles. That became my Saturday walk.

All walks provide the opportunity to more closely experience the world around you and to ruminate a bit on the days and weeks ahead. But in my case at least, some of the joy of walking around town is lessened by the constant thrum of cars driving by and the dance we have to play with them on sidewalk-less streets. In truth, I am learning to loathe the automobile. The Rail Trail, however, does not run alongside a road. It crosses a few not-very-busy streets, but otherwise runs straight and level and unmolested as the land undulates alongside it, sometimes rising above the trail and other times dipping far below. The only time the roar of cars becomes annoying is when you're crossing the bridge, but that's compensated for by the beautiful view of Newburyport laid out below you. Henry David Thoreau's description is accurate even today:

> *From the steeples of Newburyport you may review this river stretching far up into the country, with many a white sail glancing over it like an inland sea, and behold, as one wrote who was born on its head-waters, "Down out at its mouth, the dark inky main blending with the blue above. Plum*

Island, its sand ridges scalloping along the horizon like the sea-serpent, and the distant outline broken by many a tall ship, leaning, still, against the sky."

These almost-nature walks really fill your senses. And while a walk around town is lovely, a walk through the salt marsh will truly open your head. Rumination becomes full-on woolgathering as your stress levels plummet and a sense of bonhomie swells in your chest. Etcetera. Etcetera. I could go on with the whole "what a joy it is to be alive" thing, but that's rather trite and not very interesting. The point is, in case you are still sitting on the couch, even simple exercise, as simple as a walk, is pleasant. Nothing spiritual. Nothing life changing. Just pleasant.

Which is a real kick in the ass. Back in the smoking chapter, it was fun to rail against the tobacco industry. Yes, I and every other half-witted smoker on the planet knew that cigarettes were gravely unhealthy. Even so, we could still convince ourselves that the tobacco industry deserved at least some of the blame. Similarly, we can throw shade at the processed food industry that is slowly poisoning us, even though it's no secret that Hot Pockets are "foods" best avoided. But when it comes to not exercising, we have no one but ourselves to blame. Yes, today's lifestyle can make finding time to exercise difficult, but we could get up an hour earlier if we made the effort. This time there is no corporate titan steepling his fingers and coming up with the next way to separate you from your money, even if that means almost-but-not-quite killing you. Quite the opposite, in fact. There's an $80 billion industry doing its level best to get you into its gyms and wearing its sneakers. Face it, when it comes to lack of exercise, you'll find the bad guy in the mirror.

Cardiovascular Exercise

Like a well-made car, the human body can run well for quite some time, even if you ignore the maintenance schedule. But eventually abuse and neglect will catch up with you, and the system will start to fail. Just as your Miata (not MX-5, dammit!) needs regular oil changes, air filter changes, and tire rotations, your body also has lots of parts that need attention. Fortunately, you can tend to them all with just four simple actions: Eat modest portions of real food, don't smoke, do some resistance exercise, and do some cardio exercise. That's it. We've already covered the first three, so now it's time to talk about cardiovascular exercise.

For your cardio (heart) vascular (blood vessels) system to work and work well, it needs exercise, just like your muscles do. Arguably, more so. If you were somehow forced to choose between the two, choose aerobic, you'll live longer. This is because your heart is a muscle, and even though it is in constant use, pushing blood to your lungs and around your body, it needs exercise just like your skeletal muscle does. A person's average resting heart rate is about 80 beats per minute (BPM), although this varies widely among the population. Ignoring genetics, fit people have a slower heart rate, closer to 60, sometimes as low as 40 BPM, while the decidedly unfit may be hovering around 100 BPM. Taking 80 as the average and working out the math, that gives us 4,800 beats per hour, 115,200 beats per day, 42,048,000 beats per year, and 2,943,360,000 beats by the time you're 70. That sounds like a lot of work, but in a sedentary individual, the heart is not working nearly as efficiently as it could be. It's beating more frequently simply because it's pumping less blood with each contraction. In addition, as you age, the under-exercised heart and arteries start to stiffen, becoming even less efficient and generally just wearing out. The good news is, even if you get to my age, cardiac stiffness can be reversed[81] and efficiency vastly improved.

There's only one way to exercise the heart: make it work harder. You do this by exerting your skeletal muscles, the ones under your control, making them send out the "more fuel" signal. This signal is actually an increase in arterial CO_2 and a decrease in O_2. When various sensors in your arteries see that, they send a signal —*pump harder*! — to your brain, which forwards it on to your diaphragm, causing your lungs to increase the oxygen supply and your heart to deliver it.

Aerobic activity results in many positive changes to the body. First, after a few weeks of exercise, there's a persistent increase in the volume of blood, which is caused by an increase in the amount of plasma. In especially fit athletes, the absolute number of red blood cells increases too, but for the rest of us, it's mostly just the plasma increase. Nonetheless, we can still reap rewards. In particular, plasma-enriched blood is less viscous and can move more easily through the bloodstream, especially the narrow capillaries attached to the muscles. But more importantly, there's simply more blood. An increase in blood volume leads to better thermoregulation, cardiac stability, and more efficient oxygen delivery. That is, the more blood there is for the heart to pump, the more oxygen that can be delivered per stroke.

The heart, in order to satisfy your muscles' need for blood, is now working harder. We already know what happens when muscles work harder: they get bigger. The heart actually experiences hypertrophic growth due to exercise, getting larger and stronger in the same way your biceps respond to weight lifting. A stronger heart can push out more blood with each contraction, and that means it doesn't have to pump as fast to satisfy those greedy quadriceps. A decrease in

---

[81] https://www.ahajournals.org/doi/abs/10.1161/CIRCULATIONAHA.117.030617

heart rate, then, means more time for the left ventricle to fill with blood before the next contraction, providing still more blood with each push. Finally, while peer reviewed studies remain scarce, a larger heart is thought to give you the strength of ten Grinches, plus two.

In short, your heart has become a healthier, more efficient pump. After just a few weeks of aerobic exercise, both your resting and your active heart rates will decrease. Which, all other things being equal, should give you a little extra time on Spaceship Earth. According to my Fitbit data, back in June 2017, my resting heart rate was 87 BPM. It is now 72 (not including trips to the Sierras, where at 8,000 feet my heart rate increases by 10 BPM).

Cardiac health is only half of the equation; we still have the vascular bit. All that blood surging through your circulatory system, that has an impact on your physiology too. First, repeated aerobic exercise will stimulate the creation of new capillaries (angiogenesis) to help deliver blood to the muscles, something I find incredibly cool. In addition, all this extra contraction and dilation, combined with the lateral movement of blood in your veins, is essentially blood vessel exercise. And your vascular system responds to exercise much like your muscles do, by growing stronger and more supple, making blood vessels less prone to tearing, hardening, and clogging. This conditioning is not limited to just the arteries, capillaries, and veins that run from heart to muscle and back again; it extends to the entire vascular system, even the parts that feed the brain, viscera, and skin.

Ultimately, the blood reaches our muscles. Due to both strength and cardio training, muscles become super-efficient at processing oxygen and fuel. Beyond the additional capillaries now bathing the working muscle in blood, the number and activity level of the all-important mitochondria will increase. This is like dropping a V8 into your Miata[82]. Even as mitochondrial count increases by as much as 70%, so too do the number of proteins and enzymes that are also involved in cellular respiration. In addition, damaged mitochondria fuse with healthy ones, becoming larger and more powerful. Your cells can now consume considerably more oxygen and fuel than they could before, and generate much more ATP. Muscles will even get better at taking the oxygen into the mitochondria, grabbing more of it as it goes by and keeping the whole system fueled. And even though ramped up ATP production leads to an increase in reactive oxygen species (free radicals), your cells are much, much better at sweeping these up, resulting in a net decrease in cellular damage. To put all this differently, it's not so much that an active individual gains mitochondria, it's that a sedentary individual *loses* them. Since mitochondria are powering damn near everything, a loss in mitochondrial number limits the body's ability to do much

[82] Which is totally possible: https://www.flyinmiata.com/V8/index.php

of anything at all (including neutralizing radicals), resulting in much of what we recognize as being old and infirm.

You may be thinking, "Yeah, okay, fine. Now I'm more fit. Now I can do fit people things, like jogging and playing tennis. But you know what? I'm not particularly interested in jogging and playing tennis. Why should I bother?"

Setting aside the fact that you'll live a little longer, and ignoring that exercise is rather enjoyable, and disregarding that it's sometimes useful to be able to rush a bleeding toddler to the hospital, cardiovascular exercise will simply cure what ails ya. If you are not suffering from any chronic diseases, exercise will help keep it that way. If you are, then exercise will almost certainly have a profound therapeutic impact, vastly increasing your quality of life. If there's one thing your doctor and the rest of the medical community want from you, it's for you to get off your fat ass and take a walk. A lively walk.

Now, one stroll to the corner and back is obviously not going to do it. You have to get the heart pumping, and for an extended period of time. The American Heart Association recommends 150 minutes a week of moderate aerobic exercise. That's 30 minutes a day, five days a week. Easy peasy. It will take you about 30 minutes to walk 1.5 miles. That's nothing. To make sure your walk is on the spirited side, it may help to pretend there's a hand pressed against your back, pushing you along. If you graduate to running or more vigorous exercise, you can reduce the time spent exercising, not that you'd want to.

Almost immediately after starting on a (lifelong) cardio fitness routine, you will start reaping the benefits. Are you ready? The list is long. Sustained cardiovascular exercise will help prevent, manage, or eliminate: metabolic syndrome; hypertension; insulin resistance, prediabetes, and diabetes; atherosclerosis; stroke; heart disease and cardiac arrest; high cholesterol; osteoporosis; colon and breast cancer; and depression. In fact, exercise is frequently more effective at treating these chronic ailments than any corresponding drug. Moreover, a healthy cardiovascular system will boost your immune system to help fight off infection, enhance your cognitive ability, reduce stress, increase euphoria, and slow the aging process. Finally, while the jury's still out on whether cardio promotes any additional weight loss beyond the caloric burn, it is definitely the case that cardio helps you keep the weight off.

In short, exercise *is* medicine.

That's quite the litany of benefits relative to the investment, so please consider getting off the couch. You only need to keep your ears open for the signal. If your husband says he's going for a walk, join him. If you happen to wake up early on a Sunday, go for a walk by yourself. Get a dog (not a mastiff, mastiffs don't do cardio). A few weeks from now, buy yourself a Fitbit and see if you can do 10,000 steps in a day. Drop hints to your wife that maybe you'd like to take tennis lessons. Even better, dance lessons, she would love that.

Sports

In late June, Denise asked me what my sport was going to be. To which I replied with something along the lines of, "What?"

"Now that you're getting all fit and stuff," she clarified, "you should take up a sport. What do you think it should be?"

The idea of me playing a sport was ludicrous. In my entire life, I had rarely played anything more taxing than Dungeons & Dragons, and that was 30 years ago. I said as much.

"Oh, it doesn't have to be basketball or anything like that," she replied. "How about kayaking or cycling?"

"Erm, yeah, maybe. I'll think about it."

And surprisingly, I did think about it. In practical terms, there was no way for me to take up basketball or tennis or anything that requires teamwork or coordination. I'm just not skilled that way. Still, it would be nice to have a sport that I could invest time, energy, and money into. (Golf doesn't count; it provides very little in the way of cardio or strength training.) And then, sometime in July, I had occasion to visit an REI. While there, I walked past the hiking boots section. Back in the mid-90s, I and two other guys went backpacking in Yosemite for a week. It was one of the best vacations I've ever had. I'd happily do that again, and, as it turns out, we were already booked for that trip to Tahoe. I bought some boots, socks, and a pair of hiking pants.

I know I've already gone on about California, but damn, the place is beautiful. I've been to Colorado and Utah a bunch of times, Wyoming and Alberta too. Those other mountains just don't compare to the Sierra Nevada range. This is probably because I've only seen the Sierras in summer and the Rockies and the Tetons in winter, but the Sierras call to me. So, almost as soon as we landed, I went for a hike.

On this first excursion I took my younger son, Ambrose. Denise dropped us off near the trailhead, and Brosie and I walked into what appeared to be an ordinary park. Which, on closer inspection, turned out to be an ordinary park. It took us 30 minutes to circumnavigate it. Back at the so-called trailhead, Denise was long gone, and we were stuck on the Nevada side of the lake. We trudged a mile or two to the sad, ugly, and crass casinos that gather on the state line, hailed an Uber, as hearty adventurers do, and went home.

The next day, after considerably more research and closer reading, I set out again, this time with Henry. Early in the morning, we drove just a few miles to the trailhead for the Mt. Tallac trail. It's a 10-mile out-and-back trail to the summit of Mt. Tallac, 3,200 feet higher. Had my sneaker-shod son not been there, I likely would have attempted the top. As it was, we figured we'd hit the halfway mark around Cathedral Lake, then double back.

This time all went as planned. The sun was bright and the trail climbed gently into the mountains. As California's greatest poet, Dashiell Hammett, might have put it, the sky was the color blue oughta be. The high-desert flora struggled for purchase among the boulders and ponderosas. Brooks babbled and lakes twinkled. Everything was absolutely stunning, and the hike was hardly strenuous at all. It felt more like I was being carried along rather than hiking. We stopped for lunch along the lake, took some pictures, and turned back toward home[83]. It was perfect. A few days later, Henry and I drove up to Truckee, dropped Ambrose off at an indoor skatepark, and knocked off a few miles of the Pacific Crest Trail. It was just as nice. Indeed, I think I found my sport. Sunshine, no cars, friendly people, beautiful scenery. There is absolutely nothing not to like about hiking.

Of course, you also get a great CV workout from hiking, and the workout comes with a few bonus features. For one thing, hiking tends to be more difficult than walking, so you get your heart rate up higher. If there's any altitude gain at all, you also work your muscles harder. Work them hard enough and it counts as resistance training, and you get to skip leg day at the gym. When hiking, you have to frequently engage a lot of stabilizing muscles in order to navigate the uneven terrain, which should strengthen your abs. Finally, and back to my earlier point, it's just lovely to be out in the woods or the park. Sitting by an ice-blue lake or skipping easily over a rock-strewn stream is guaranteed to lower your stress levels, and that will lessen your risk of heart disease, high blood pressure, and infectious disease. But if hiking is not your thing, then do indeed try tennis or taekwondo or cycling or boxing or basketball or really just about anything that gets you moving — ideally outdoors.

---

[83] I've since returned, and, along with Denise, hiked to the top. A couple of days later, we also hiked all 13 miles of The Rubicon Trail.

Continuous Endurance Training

All too soon, I was back in Newburyport. Later that winter I would start hiking in the White Mountains of New Hampshire, but that was still months away. In the meantime, I resumed my normal walks around town. Then, around the middle of August, while on an ordinary Rail Trail walk, the damnedest thing happened. Without planning or even conscious thought, I started running. One moment I was walking along contentedly, thinking about nothing at all, and the next I was running. I doubt it was more than 100 yards, but I ran it.

After that brief spurt, I walked for a few more minutes, then I did it a second time. This time I ran until it felt too difficult to run any further, call it 200 yards. I slowed back to a walk, and kept walking until my breathing returned to normal and my legs felt unfatigued, then — Britney Spears style — I did it again. And again. I hit the turnaround point and continued to alternate between walking and running for the length of the rail trail, predominantly walking. If I had to guess, of the five miles that make up that walk, I probably ran one of them. Which doesn't suck even a little bit.

The next weekend, no longer all that innocent and sporting one of those things that allow you to strap a phone to your arm, I walked back to the Rail Trail for a repeat performance, this time trying to run as much of it as possible while listening to The Talking Heads. I think I did okay, adding at least another half mile, likely more, to the distance I ran the previous week.

In mid-September, Denise joined me for a run. That was nice. Our running styles are very different, though. I have a large stride and at the time I would run to exhaustion, then walk. Denise has a short stride and likes to run for a fixed period of time, then jog for a little, then run again. It was awkward. But not long into that run, Denise pointed out that the trail had markers every tenth of a mile, so she switched us up. Instead of running to exhaustion, we'd both run a tenth of a mile, walk a tenth of a mile, run a tenth, etc. Which is a great way for a beginner to take up running. Just as you feel you're getting tired, you get to walk, and just as you feel the desire to run again, you can. I was later to learn that this is called interval training (or walk-run), and it's a very enjoyable way to run, even if tenth-of-a-mile increments are on the short side.

We repeated that run the next week. But the week following that, now that we could run together, Denise took me to a favorite running path of hers. We had to share it with some cars, but there weren't many. Denise wore a Nike-branded Timex that she configured to chirp after 60 seconds, then again after two minutes, and repeat. We used that in lieu of the markers. We would run for a minute, walk for two, then run again, and so on. It was very nice. The jog took us past some attractive homes, down along the salt marsh, and around some parkland. The whole way Denise had to put up with my incessant chatter: "It's nice back here. I wonder how many other people know about this? Do you run here a lot? Are there special techniques to running? The air smells so clean. This is fun. Do you have other routes like this? I bet Iggy Pop is good running music[84]."

After a while, Denise laughed and said, "I finally figured it out. You are experiencing for the first time what I've been telling you for years, you lunkhead; running is fun." She was right, running is fun. At least the kind of light interval running we were doing. It was nice to be outside on a lovely day, and it was very likely that there was a small endorphin rush going on. And then she said, "I bet you're kicking yourself for missing out all these years."

Pow! Right in the feels. She was right again. Thirty-seven years of smoking and a lifetime of sloth. What a fucking waste. But now's not the time for brooding.

The interesting thing about the decimile (which I'm pretty sure is a word) markers and Denise's watch is that I seem less inclined to slack off if something, if only a signpost or a faint beep, is telling me to move. This effect is even more pronounced if a human (-like) voice is telling me to go "steady for two minutes." So I waded through a morass of mobile apps designed for runners, and ended up downloading *Runkeeper*, which has a variety of interval presets, route mapping, and most importantly, a computerized voice telling me to go slow or go steady.

In the meantime, Denise had signed us up for the annual Thanksgiving Day Turkey Trot, a 5K race through Maudslay State Park, which is a quaint 500-acre park near the edge of town. In late October, I got myself over to Maudslay, strapped on my phone, and went for my first trail run, which was just deliriously fun. I absolutely had the wrong shoes on and I nearly twisted my ankle on a leaf-covered rock. But it was a joy. Very much like hiking, only faster.

I went back to Maudslay about three more times before the race, each time changing the interval settings in *Runkeeper*. I went from one minute running and two minutes walking to almost the exact opposite, two and a half minutes running and one minute walking. On Thanksgiving morning, Denise, Henry, myself, and seemingly ten thousand other people milled around the great field ahead of the race. Ultimately, something caused the crowd to surge forward, so I surged too. I fumbled with my phone, trying to start the perfectly curated running playlist I had made the day before, and got separated from Denise and Henry almost immediately. In the end, I finished the race in 45 minutes and 34 seconds, which anyone else would consider a pretty crappy time, but I was pleased with it — even if my 13 year old son beat me to the finish.

---

[84] Iggy Pop is excellent running music.

For the next few weeks, I found other fun places to run: around the reservoir (3.4 miles), on the new extensions to the Rail Trail (3 miles), and once in New York City, from the Javits Center on 34th St down to the Staten Island Ferry (4.2 miles). I even switched my weekday walks over to interval running. According to *Runkeeper*, my fastest 5K as of July 2018 took 30:41 to complete, nearly 15 minutes faster than my first!

So, contrary to all expectations, I am a runner. Well, not a runner the way other people are runners. Let's just say, I enjoy running. Not far and not hard, but often. It's not painful and it's not boring. For everyday continuous endurance training (the fancy term for the somewhat leisurely cardio exercises that primarily leverage your slow twitch, type I muscles), I still prefer three-mile interval-style runs, ideally on a wooded trail, and I probably always will. All in all, running is now part of who I am, and the next time someone says they're going for a run, I'll ask if I can join them.

High Intensity Interval Training

But back in early July, when running defined me about as much as line dancing did (which is to say, not at all), I only had my walks. As pleasant as those walks were, they weren't quite the boot camp replacement I wanted. Even for me, they were a little too serene. I started looking around for something I could slot into the Tuesday/Thursday gap to really rev the old ticker. This was when I learned that "high-intensity interval training," HIIT, is a real thing, and not just the name of one of Fuel's cardio offerings.

HIIT is basically what it says on the tin: a cardiovascular exercise plan characterized by short spurts of high-intensity effort, followed by a brief interval of low-intensity effort, repeated several times. A common interval is 20 seconds of high-intensity anaerobic exercise, followed by 10 seconds of low-intensity aerobic exercise. Do, say, three sets of six reps with a minute in between, for a total of 11 minutes, and you have (an example of) high-intensity interval training.

What exercises are performed, how long each on/off period is, how many sets, the time between sets, and basically every other factor can and does vary greatly. The point is, if you really dial up the "H" in HIIT, a super fit athlete can get a great CV workout in four minutes that's just as effective as if they had gone for a 60-minute run. Even ordinary folk can complete a cardio workout in less than 30 minutes, and that's pretty cool. Except for one thing. By "high-intensity," they mean as hard as you can go, or more realistically, 90% of what you're capable of. While the magazines might say, "You can do anything for 20 seconds," the fact is doing anything at 90% for 20 seconds, and then doing it again, and again, and again, well, it's kind of grueling. And that can make HIIT the kind of exercise that is more honored in the breach than the observance.

Does it work? Absolutely! A HIIT-based cardio plan is every bit as good as its more time-consuming but less arduous counterpart, continuous endurance training, CET. Is it better? No, it's not better. Study after study has concluded that both types of cardio are equally effective. It's true, there are some differences — CET will develop your heart better while HIIT will burn more fat — but they're minor. The principal difference is that HIIT is quick and CET is fun. Despite what the magazines say about anaerobic vs aerobic, and despite what the websites say about shredding it for summer, you would be equally well off doing either. That said, it may be best to do a little of both. Take a little less time to burn a little extra fat and take a little more time to lower the stress levels. That's where I ended up.

It took a bit of fumbling around, but by mid-July I had zeroed in on what would become my regular HIIT plan. It's a very slightly modified version of a beginner plan I found on the Daily Burn website[85]. Perform each of the following exercises as hard as you can for 45 seconds, take a 15-second break, then move on to the next. Repeat three times, resting for 60 seconds between sets.

1. Squats
2. Push-ups
3. Butt kicks
4. Tricep dips
5. Crunches
6. Jumping jacks

When I do this routine, I hit it pretty hard — and this next modifying clause is important — for me. I do *not* have the intensity dialed up to eleven. I couldn't sustain these exercises for 45 seconds if I did. Also, there's kind of an upper limit on exactly how fast I can do a push-up. Still, I'm working it, and after the very first set of squats, I'm already breathing heavy. By the time I'm doing the butt kicks, I'm sweating. But to be honest, my intensity is more medium-high than high, maybe 75% of max. Which is okay by me; it makes this routine more enjoyable and thus less likely for me to quit. And, importantly, it's still a very effective workout that takes *exactly* 20 minutes!

[85] https://dailyburn.com/life/db/hiit-workouts-for-beginners/

That Song by The Clash

By late July, I had arranged my morning workouts such that on Monday/Wednesday/Friday I would go up to the attic and do the HIIT cardio, and on Tuesday/Thursday/Saturday I would do the CET cardio, walking and/or running. On Sundays I continued to do nothing at all. Doing HIIT (okay, MHIIT) exercises in the house kind of sucks. There's limited room, you get all sweaty, and you may find yourself bouncing up and down on the ceiling above your sleeping family's heads. But I had no choice, I don't have a yard and I wasn't about to do this in public. God forbid!

Then one day in September, I had to skip my morning cardio. No biggie. It was a Wednesday, which meant I wouldn't be at the gym lifting that night, so I could just make up the cardio then. Except Henry, my burgeoning metrosexual son, had to make an appointment with the only person in a 30-mile radius who is capable of styling his hair. But I am a Pretty Good Dad, so I changed into my workout clothes, drove to Amesbury, dropped off the kid, and went looking for a place to do my HIIT routine. Outside! In public!

I knew there was a postage stamp-sized park a couple of blocks from the stylist, and I headed that way. This took me past a low-slung building with wide doors open to the world. I slowed as I could see it was some kind of gym. Thinking that I might be able to pay a drop-in fee and not have to do my cardio outside after all, I stepped through the doorway. Inside was a wide space. It was difficult to see, as my eyes were still adjusting from the bright sunlight, but there were ropes hanging from the ceiling and barbells strewn across the floor. A group of strikingly fit people were standing in a semicircle, facing someone who was obviously their officiate.

One of them noticed me. "Come on in," she said, smiling. She was exquisite. Not an ounce of fat on her. Her tummy as tight as a snare drum. Her muscles taut under tan, glistening skin.

"I'm — I'm not a member," I stammered back. *Had she seen my workout clothes and assumed I was one of them? How could she? I could never be that strong.*

Alerted to the scent of new blood, the others turned to me in unison, their flaxen hair framing chiseled features, their eyes taking on an eldritch glow. "Don't worry," they said as one, hands reaching out beseechingly. "Come try it out. It's fun."

I took a step backward, suddenly afraid. "It's okay. Some other time. I've got — I've got to go." I turned and ran.

Later, when I relayed these harrowing events to Denise, she informed me that it was a CrossFit studio and that I am not allowed to do CrossFit as I am insufferable enough already.

But I might.

I might.

They were so very beautiful.

I did learn one thing that day: calisthenics outside is no more embarrassing than running outside. So instead of HIIT in the attic, I could do it in the park that's right across the street from my house, the one with the statue of George Washington. This park also has the advantage of having a pond in the middle of it. In fact, the park is pretty much just the pond with a walking path around it. The pond happens to be sunken about 20 feet below street level, which means I could do my exercises on the edge of the pond, far from the prying eyes of everyone except the soon-to-be-inspired (or so I imagined) people walking their dogs.

Starting in early autumn, on every other weekday morning, I'd grab my phone, Bluetooth speaker, and yoga mat, and walk across the street. I'd set up right by a bench that sits next to the pond, a bench I could use for tricep dips, incline and decline push-ups, and as a place to rest the speaker. Then I'd work through the routine outlined above while listening to Nick Lowe and waiting for my future acolytes to discover me. Some days Denise would join me, most days she would do yoga or barre instead. It was quick and enjoyable, even on the cool mornings and even as it got darker and darker — maybe especially then.

By the time an unseasonably warm November rolled around, I had things pretty well in hand. I'd get up at 5:30 every weekday. On Mondays, Wednesdays, and Fridays, I'd do my HIIT routine. This meant I was back in the house by 6:00, early enough to shower and dress before waking the kids for school at 6:30. I would then have a large and leisurely breakfast, and leave for work at 7:30, maybe 8:00 if I felt like it. On Tuesdays and Thursdays, I'd do the three-mile run/walk around the still-under-development Rail Trail, returning home by 6:30. On Saturdays, I'd go for a trail run through Maudslay or a street run around the reservoir.

My Fitbit tells me that between September 1 and November 30, I was burning 3,200 calories a day on average and I walked or ran some 6,000 miles. Then December came and shot everything to pieces. It got cold, it snowed and rained and slushed, then it got really cold. I mean really fucking cold for a really long time. I soldiered on for the first few weeks, but by Christmas my cardio plan was reduced to tacking an extra 30 minutes on the treadmill to my strength training sessions at the gym. In January, I managed to get some hikes in, and in February, a few street runs. But three successive Nor'easters (which used to describe a very particular type of storm, but now just means blizzard) in March pretty much eliminated any choice but the treadmill, and I hate running on treadmills.

Flip Flop, Fit to Drop

Tuesday, July 31, 2018. Seven months later.

Seriously? A hill? They put the finish line at the top of a goddamn hill?

Fortunately, St. Steven d'Cupertino, the patron saint of shuffle play, has decided to end my run with *Rip this Joint* by the Rolling Stones, and that gives me the strength to slog forward. Nevertheless, I swear out loud at the strain the hill necessitates. This causes the people gathered on the sidelines to laugh. I can hear them even over the music playing in my ears. But they mean well, they clap and cheer me on. It's not much of a hill, really. Can't be more than twenty-five yards to the top, where, outside the front doors of the high school, it angles back down again for another twenty-five yards to the finish line. As I crest the hill, I hear the announcer call my name: "And here comes number 1024, Peter Lacey. Another Newburyporter[86]." I raise my fist, Tommie Smith style, as I cover the last few yards. Then I'm over the line and thinking, *I forgot to look at my fucking time.*

It was 1:44:31.

By any objective measure taking nearly two hours to complete a ten-mile race is pretty poor. Overall, I placed 797th out of 946 — the 14th percentile. In my age group, men 50-54, I placed 68 out of 73. Those are disastrous numbers. But since I'm talking about myself, I am under no compunction whatsoever to be objective. In fact, I can be as subjective as I damn well please. And subjectively, I rock! Denise, who also ran the race, she rocks too. In fact, all of those 149 racers who finished behind me, they also rock. So do all the runners who placed ahead of me. Everyone rocks! And we rock because we ran a 10-mile race in the early evening of the last day of July during one of the hottest, most humid summers to date.

Truth be told, though, I may rock just a tiny bit more than the others. That's because a year prior to this race I was smoking a pack of cigarettes a day. A year prior to this race I was 30 pounds overweight. A year prior to this race I couldn't run a block, let alone ten miles. Also, had it not been for the heat, I think I could

[86] Denise prefers the term "Newburyportion."

have done a lot better. Just two weeks earlier, during training, I ran nine miles at a pace of 9:49 per mile — in the rain. If I was able to maintain that pace on race day, I would have finished in 678th place. So, yeah, I think I rock. Subjectively. All things considered. For a certain definition of "rock."

Four months earlier, when spring finally arrived, Denise and I had thrown ourselves into running with renewed vigor. For no good reason, we started doing long runs every Saturday. By the middle of May 2018, we completed an eight-mile run.

"You know there's a ten-mile race during Yankee Homecoming, right?" Denise asked me one evening. Yankee Homecoming is Newburyport's annual weeklong celebration of itself, complete with parades, classic car shows, bed races, fireworks, and the annual Lions Club 10-mile race. Yankee Homecoming is also at the end of July, when it was likely to be hot and sticky.

"I am so aware," I answered.

"Are you doing it?"

"Maybe."

Then the people at Fuel sent out an email, offering an eight-week training program to get ready for the race. The email arrived while I was at work and gnawed at me the rest of the day. Later that afternoon, I picked up my phone and registered. A few days later, so did Denise.

And so, on June 2, 2018, a little more than a year after Denise and I had gathered by the river for a six-week boot camp, we gathered at the Fuel Training Studio for an eight-week running camp. We arrived at 7:00 AM, and even though summer wasn't officially here yet, it was hot and muggy, a harbinger of the torrid weather to come. It was a small group, just nine of us, me and eight women[87], not including our running coach and force of nature, Meagan.

Slender and energetic, as you might expect, Meagan took the ragtag group in front of her and patiently shaped us into disciplined runners. With kindness and respect, she prepared each of us for the road ahead. She helped us with our form (don't clench your fists), she offered advice on how to deal with hills (chin up, lean forward slightly, keep your eyes on the horizon), she even taught us how to psych ourselves up for a long run by repeating a mantra in your head, such as, "I am trained. I am strong. I am ready for this."

Most impressive of all, however, was how she managed to spend time with each of us on every run. The group was of varying ability: two of the members could run circles around the rest of us, I was next in ability, with Denise after me.

---

[87] Single men, are you paying attention?

I'm not sure how the others compared, but on any given day, there was at least a mile, maybe two, separating the front runner from the back of the pack. Even so, like a bee in a flower-dappled field, Meagan would flit back and forth among us all, doling out encouragement, advice, and bottled water. She was indefatigable.

The plan was to meet every Saturday morning for the next eight weeks, each week adding one mile to the length of our runs. We all received a schedule that had us supplementing the Saturday runs with runs on Tuesday and Thursday too, and with strength training on Monday, Wednesday, and Friday. In other words, I was given a training schedule that was almost identical to my already existing workout schedule. The weekday runs varied slightly from each other. They were mostly three and four mile runs, though there was one six-miler. We were also to change the tempo on some runs, and find some hills for others.

But the Saturday runs were the real training. These all took place on some subset of the actual course, which itself was a superset of the three miles I typically ran anyway; that is, they were in-town street runs not far from my house. I won't bore you describing each of them; mostly it was just a matter of getting used to running long distances and that damn hill at the end of Merrimac St. that continues onto Spofford. God, I hate that hill.

The night before the race, I went to bed praying that someone would restart the jet stream and send a high-pressure system our way, but race day dawned hot and humid, as it had every previous day for the last four weeks. As instructed, Denise and I ate our typical breakfast, in my case, a bit of protein (steak) from the night before and two eggs. And we ate our typical lunch, in my case, the rest of the previous night's steak. At 4:00 PM, we were instructed to have a carb and protein-rich snack; I had a peanut butter and jelly sandwich. We were also instructed to avoid dairy, so I washed the PBJ down with a glass of milk because I'm not a Philistine. At 4:30, Denise and I walked a mile to the starting line at the high school. The race was actually two races in one, a 5K and the 10-miler. Consequently, there were approximately 2,000 people milling around.

A few minutes after the 5K runners set out, the 10-milers gathered at the starting line. At 6:10, some signal must have gone off, because everyone started running. For a half mile or so, while the course was still crowded, Denise and I stayed together, but as runners spread out according to ability, we gradually separated. I ran past my house, but my children were nowhere to be seen — the ungrateful little shits.

The first few miles were fine. My pace was intentionally slow as I ran through every sprinkler and accepted every proffered cup of water (as instructed). Around the four-mile mark I felt the need to pee, and that kind of sucked. Fortunately, I knew that in about a mile we would be passing the Little League ballfields with their bathrooms. At mile five, I separated from the pack, ran into the thankfully empty restroom, and relieved myself.

By mile six, the humidity was getting to me, so I allowed myself a ten second walk. Shortly thereafter we hit the hill on Merrimac. I ran/walked it. Next came the even steeper hill on Spofford. I ran/walked that one too. My overall time was collapsing as we turned onto Ferry Road for the second leg of the race. I reached for my mantra:

I must not fear.

Fear is the mind-killer.

Fear is the little-death that brings total obliteration.

But by then I was already done. The heat and humidity had taken its toll, and even the Bene Gesserit *Litany Against Fear* couldn't help me. I ran/walked most of the next four miles, running slowly about 75% of the time and walking the rest. I did run the entirety of the last mile, however, just because it would have been lame not to.

One hour and forty-four minutes after starting the race, I crossed the finish line. My quads and my groin were aching, as were the bottom of my feet. I was breathing heavy, but my lungs were good. Sweat was dripping out of every pore[88]. I wandered around in a daze for a few minutes before remembering to go up to the barriers and wait for Denise. Some minutes later, the announcer called out Denise's name as she came over the top of the hill. Moments after that we closed the distance between us, grinning like two programmers who just got their C code to compile without errors or warnings. We hugged and kissed and congratulated ourselves, then we turned and limped toward home.

[88] The White Stripes are also on my running playlist.

Chapter 8: Blaze

The last half mile of the trail, from the Sandwich Dome fork to the trailhead, trends slightly downhill. I take it at a trot, my microspikes digging into the snow and ice, my breath periodically clouding the air in front of me. It's not easy running in hiking boots and spikes on uneven ground, and I'm stabilizing cores I didn't even know I had. A hiker heading the other way stands to the side of the trail to let me pass. "Having fun?" he quips as I go by. "Absolutely," I reply. And I am. I'm having a blast.

The giddiness associated with exercise, especially outdoor aerobic exercise, is well documented but came as a huge surprise to me. And with each new thing I try, it surprises me all the more. HIIT, jogging, trail running, hiking — especially hiking — all of them make me feel so good that I start to get a little punchy when I can't get outside. Even if a run isn't that great for whatever reason, it's still better than sitting inside, and I feel better for having done it. It's why the winter of 2017-2018 — especially the six weeks after Christmas and all of March — sucked so much. It was, in turn, brutally cold, unseasonably warm, then uncommonly snowy, so much so that I barely got outside at all. I used to think that "cabin fever" was a figurative term. I can tell you now that it is quite literal.

Prior to boot camp and my first run, I had always envisioned exercise as uncomfortable if not outright painful, something more to be endured than enjoyed. I just assumed that the runners on High St. were out there running because they were better people than I was, and were pushing through the discomfort in order to reap the rewards of being fit. That seeming tradeoff never really worked for me (or so I told myself); what's the difference between suffering now and suffering later? Why not smoke and eat cookies and be lazy? It's a lot more enjoyable than shin splints and side stitches. I knew I was rationalizing, but still, I never imagined that people were outside running because they really *needed* to be outside running. And they — hell, *we* — need to be outside running, because running makes us feel good. It makes us happy. But why is that?

◐ ◆ ◑

One of the more intriguing theories as to why humans lost their body hair and evolved an extraordinarily efficient cooling system is that it made us better runners, and therefore better hunters. Imagine a tribe of early humans looking to bring home the carcass of some calorie-rich ruminant. They have no projectile weapons to speak of. Heck, they only recently figured out how to sharpen a stick. And right over there, down by the waterhole, is an antelope, strong and graceful and just bursting with protein and fat. The problem for the humans is that the antelope can easily outrun them. The problem for the antelope is that it can't do so for long.

All this ancestral pack of humans needs to do is spook the beast. It will run away, maybe even hitting its top speed of roughly 50 MPH, but it will stop once it thinks the danger has passed, panting a little and looking around for water. Until, that is, those damn humans show up again, and off it will go at a trot. By the time it stops next, it will be breathing heavier and be much warmer. Then those hairless apes will come at it one more time. And they will do it again and again, relentlessly chasing the animal down until, hours and miles later, it collapses to the ground, suffering from severe heat exhaustion. At which time, one of the humans will simply walk up to its prostrate form and stab it in the heart — and he won't even be breathing hard.

The fact is, a fit human is capable of running down any animal on Earth, even horses and dogs, the animal kingdom's best long-distance runners. A horse can probably trot continuously for 15 miles before exhaustion sets in, less if the weather is too warm. In contrast, the American ultra-marathoner, Dean Karnazes has run 350 miles without stopping, and he did this in just 80 hours and 44 minutes, averaging 13 minutes and 49 seconds per mile[89]. The ability to perform sustained cardiovascular exercise isn't limited to elite runners and hunter-gatherers; it's what each and every one of us have evolved to do. The problem is, for most of the world today, there's very little reason to run down an antelope, chop down a tree, or even lift a bale of hay. In the Western world, most of us are sitting behind desks, standing behind counters, or getting behind the wheel.

And so we exercise. When we do, whether it's trail running in the winter, 10-mile races in the summer, or hiking in the fall, we call upon all the same systems that our antelope-slaughtering forbearers did. Most of these systems react to stress. Not just stress in the sense of danger, but also in the sense of physical exertion. In the same way that your heart and lungs and muscles and blood vessels

---

[89] To put that in perspective, that's roughly the distance from Edoras to Minas Tirith. Meaning Karnazes could have reached Denethor in only twice the time it took Shadowfax.

respond positively to exercise, so too do the systems that regulate our bodies. These systems have to control temperature, maintain oxygen levels, deliver fuel, remove waste products, and dampen pain. Maybe even make us enjoy the hunt. Exercise these systems regularly, and they will get better at doing their jobs. Fail to do so and you will feel stressed, listless, and out of sorts.

To maintain homeostasis, especially under extreme conditions, all of these physiological systems need to be working in perfect concert with each other, and so we have evolved a variety of signalling hormones and neurotransmitters to coordinate these various functions. You have some level of familiarity with these chemicals already, whether that be from cocktail parties, the news, or even earlier in this book, so we won't dive deep, but let's take a peek at how all these functions are modulated by the nervous system and endocrine system (a collection of hormone secreting glands located throughout your body).

Endocrinology and Neurochemistry Part II

First up, norepinephrine, aka noradrenaline, and its close cousin, epinephrine, aka adrenaline[90]. These are the action verbs of your endogenous pharmacopeia. They are here to get things done, whether that be fighting or fleeing, and they are going to direct every resource at their disposal to better allow you to do it. Since exercise is interpreted as stress, these and other stress-related messengers are put into play almost as soon as you start to move. Interestingly, these two chemicals are simultaneously neurotransmitters and hormones. Norepinephrine is primarily, but not solely, a neurotransmitter, and epinephrine is primarily, but not solely, a hormone. They each can do the other's job, but they do specialize.

Let's say you're laying siege to a French castle, as one does, when some "nimble gunner with linstock now the devilish cannon touches." Your eyes see the flash, your ears hear the boom, and an electrochemical wave — an action potential — is sent along your central nervous system, from your brain and down the long bundles of neurons, i.e. nerve cells, in your spine. This triggers these neurons to release the neurotransmitter acetylcholine, ACh, which you may remember from the discussion on nicotine addiction. The ACh jumps across synapses in the multitude of places where the central nervous system is wired to the sympathetic nervous system (SNS). The SNS is the part of your autonomic nervous system that's responsible for fight or flight responses, and therefore

[90] The names are perfect synonyms, it's just that the name Adrenaline was originally a trade name. Epinephrine is the more "scientific" name.

connects to pretty much all of your important bits, e.g. muscles, heart, lungs, blood vessels, genitalia, various glands, even your eyes.

When the ACh triggers an action potential in the nerves of the SNS, the so-called postganglionic fibers, it causes them to release a different neurotransmitter, norepinephrine. And when the norepinephrine starts flowing, your important bits just go nuts. Norepinephrine makes the heart beat faster and constrict more powerfully; it increases blood pressure and thus blood flow to the muscles, while decreasing flow to the skin and digestive system and other non-survival-related metabolic processes; it promotes glucose uptake by muscles; it evens makes your vision more acute. Furthermore, a small amount of norepinephrine is also released in the brain, primarily making you more alert and attentive.

Your sympathetic nervous system also has a direct connection to your adrenal glands that are resting on top of your kidneys. More specifically, it is connected to a subcomponent deep inside the adrenals known as the adrenal medulla. The adrenal medulla is responsible for synthesizing both epinephrine and norepinephrine, and when it gets jolted by the SNS, it releases both of these hormones into the bloodstream. Yes, because this batch of norepinephrine is in the bloodstream, it's now considered a hormone and not a neurotransmitter. It's not a 50/50 mix, however, it's closer to 85% epinephrine and 15% norepinephrine.

Conveniently, the bloodstream can reach where the SNS can't, although it's a little slower getting there, allowing your muscles and adipose tissue and liver and pancreas to bathe in a rich cocktail of hormones. In the blood, epinephrine performs all the same jobs as norepinephrine, including inhibiting the release of insulin while triggering the release of glucagon, both actions signalling the liver to break down glycogen (glycogenolysis) and to synthesize glucose (gluconeogenesis), releasing the resulting monosaccharides into the bloodstream. In addition, epinephrine directly triggers glycogenolysis in muscle cells as well as the liver. Furthermore, both glucagon and epinephrine increase lipolysis in fat cells and stimulate the release of fatty acids. In other words, adrenaline is asking for all the fuel it can get — fat, glucose, ketones, everything — and priming the muscle cells to take it all in. All of that fuel will need to be oxidized, of course. Accordingly, norepinephrine also dilates the bronchioles in the lungs while increasing the breathing rate, thus allowing more oxygen into the bloodstream and into those ATP-hungry muscles.

Since most people are familiar with it, I should also mention cortisol, a hormone very similar to adrenaline in both makeup and effect. Cortisol is normally released by your adrenal glands throughout the day, and has a large number of regulatory roles to play, including influencing your metabolism and controlling blood sugar levels. Extra cortisol is released into the bloodstream if the body senses that it is under prolonged stress, that is, if the stress is chronic and not acute. It follows after and lasts longer than adrenaline, while doing mostly the same job. Not including slow-burn, twenty-first century stressors, such as commutes, deadlines, and mortgage payments, excess cortisol is mostly evident only during especially strenuous exercise.

In short, these three chemicals — norepinephrine, epinephrine, and to a lesser extent, cortisol — give you a jolt of energy and open up all the chemical pathways available so that you can use it. When Henry V roused his men to "imitate the action of the tiger; stiffen the sinews, summon up the blood, and disguise fair nature with hard-favour'd rage," he was very consciously eliciting a response from their hypothalamus and adrenal glands.

It's interesting to note that epinephrine is made from norepinephrine, and norepinephrine is made from dopamine. You remember dopamine, right? It's the primary driver of the "seeking" part of your reward system, and the primary driver of addiction. When you exercise, your brain starts pumping out the stuff. Dopamine, recall, makes you want something. In the worst case, it may make you want a cigarette. In the best case, it makes you want to dance or run or hike or bike or swim. It especially wants you to achieve a goal, like completing a ten-mile race or running your fastest 5K. Dopamine is why I need to go outside most every day to walk or run or do some interval training. I need it in much the same way I used to need a smoke. I won't go as far as claiming that fit people are addicted to exercise-induced endogenous dopamine, although some are, and the rest don't mind it.

Also, like cigarettes, exercise (or antelope hunting) triggers an endorphin response. Endorphins, you'll recall, are the "consummating" part of the reward system. Endorphins are interesting because they are not released immediately during exercise. In fact, they're not released until your body starts to flag, and then your brain and nervous system open the taps, crying, "Now set the teeth and stretch the nostril wide, Hold hard the breath and bend up every spirit." By binding with the opioid receptors of your brain, endorphins do more than just light up the hedonic hotspots, they play the now very important role of blocking pain and controlling emotions. That is, endorphins *are* opioids. They're endogenous painkillers, little different from exogenous morphine. In fact, the very name "endorphin" is a portmanteau of "endogenous morphine." Popular culture still has it that endorphins are responsible for the "runner's high," but more and more it seems that they are not, at least not entirely. The principal driver of that particular sensation of euphoria is our next chemical.

Endocannabinoids, as you likely guessed from the name, are neurotransmitters that bind with cannabinoid receptors located throughout the body. If endorphins are smack, then endocannabinoids are pot. The endocannabinoid system is bewilderingly complex, and responsible to one degree or another for modulating a great deal of behavioral and psychological effects. In relation to physical exercise, the endocannabinoids are there to mellow you out, reduce pain, and lower anxiety. Unfortunately, like with endorphins, your body's not going to get you high until you need it. I'm not sure exactly how far or how hard you have to run to trigger a runner's high, but I can tell you it is frustratingly *not* ten miles at 10:27 per mile.

So far, we have the stress response chemicals, norepinephrine, epinephrine, and cortisol, priming us physiologically. We also have the reward center neurotransmitters, dopamine and endorphin, letting us enjoy the hunt and suppressing any pain. If we need an even bigger hit than endorphins can give us, the endocannabinoids will get us over the hump. But all of these are of the moment. When the hunt/fight/run is over, they all start to dissipate. However, there are exercise-stimulated chemicals that seem to have a more lasting impact on our mood and behavior. Chemicals which will not only help you bring down the antelope, but enjoy the antelope stew afterward.

Serotonin is a neurotransmitter that your body synthesizes from the amino acid tryptophan. Yes, that tryptophan, the stuff in your Thanksgiving turkey — which would be present in your holiday dinner at the same level if you had a Thanksgiving burger instead. A small amount of serotonin is continuously manufactured and released in the brain, and exercise will, of course, stimulate the release of even more. In the brain, serotonin helps modulate motor coordination, appetite, sleep cycles, and pain management. Serotonin also dampens dopamine production. The evolutionary idea behind that being that dopamine makes you want something (for example, food) and serotonin tells dopamine to shut up so you can enjoy it. Most importantly, the effects of serotonin in the brain linger — for a long time. Serotonin, then, is more associated with mood and personality than pleasure and pain. People with high levels of serotonin tend to be social, agreeable, and in group leadership positions. Moreover, serotonin helps you feel good about yourself, allowing you to feel confident and successful. Long term, this can lead to lasting accomplishment and contentment. Of course, there's a flip side to this. Too little serotonin is suggestive of depression, anxiety, aggression, neuroticism, and suicide.

Similar to serotonin is oxytocin. You may know of oxytocin as the "cuddle drug" or the "love potion," as it is directly associated with romantic love (not lust, that's dopamine), parental nurturing, friendship, trust, loyalty, and more. As with other neurotransmitters, physical exertion is not the only oxytocin trigger; others include yoga and meditation, foreplay and orgasm, reading and good music — in case you're looking for something to do this weekend. It may seem odd for evolution to express oxytocin during exercise, but it makes sense when you realize that hunting and warfare (another thing we as a species are particularly good at it) are group activities demanding of cooperation, trust, and loyalty. When Henry exhorted his dear friends to go once more unto the breach, he was not just prodding his men into battle, he was asking them to die for king and country, and they were probably up to their eyeballs with oxytocin. Another win for Henry is that oxytocin only bestows these positive emotions on members of the "in group." When dealing with the "other," in this case, the French, oxytocin elicits a strong aggressive response.

Exercise also promotes the release of two more neurotransmitters: glutamate and gamma-aminobutyric acid, or GABA. Glutamate is a morning person. It's excitatory. That is, it promotes neuronal activity. During exercise (hunting, fighting), glutamate is putting your body under stress, making sure everything is ready for whatever comes next. It makes you pay attention and helps you remember what you learn. GABA is the exact opposite, it's inhibitory. It just wants everyone to calm the fuck down. Like serotonin, it inhibits dopamine production, and during exercise, GABA lowers stress and anxiety significantly.

By the way, if you want to lower stress and anxiety, but don't feel like going for a run, then you should know that yoga also promotes GABA production — in everyone but me. Besides my complete lack of grace, there may be another reason why I don't find yoga relaxing; my drinking. It turns out that alcohol binds to GABA receptors, which is why most people start to feel relaxed after a drink or two. Excessive drinking, however, can damage GABA receptors, just as other addictions can damage dopamine receptors, making it increasingly more difficult to lower stress levels and encouraging still more drinking. That said, I don't think that my level of alcohol consumption has done any significant damage here, it's more likely that I'm just ungainly.

Lastly, I'll touch on one lesser-known protein expressed during (vigorous) exercise, but which has nothing to do with your immediate happiness or stress relief. It is called brain-derived neurotrophic factor (BDNF), and its job is to promote neurogenesis, the creation of brain cells. You may be thinking that brain cells don't regrow, and largely you're right; most of the neurons you have now, you've had all your life. But some are continually refreshed, especially those associated with learning and memory. And the principal modulator of neurogenesis is BDNF. I mention it here because BDNF levels are directly associated with a number of degenerative brain diseases, such as Parkinson's disease, multiple sclerosis, Huntington's disease, and (particularly worrying to me — stay tuned) Alzheimer's disease. As with oxytocin and GABA, yoga and meditation both stimulate BDNF production.

As earlier mentioned, there are well over 100 neurotransmitters and hormones that alter the behavioral, physiological, and psychological properties of the brain and body. The brief tour above was both superficial and reductionist. But the fact is, all this neurochemistry is mind-numbingly complex and infinitely variable. In the large, the lesson to take home is that exercise is both pleasurable and a genuine source of happiness. And as you likely know, independent of any of the direct health benefits of exercise, just being happy makes you healthy.

Happy

It's well known that continuously subjecting your body to stress and anxiety, be it fear, violence, poverty, or other similar experiences, has a profound negative impact on one's health. Stress, and the corresponding oversupply of cortisol, is strongly correlated with stroke, cardiac arrest, high blood pressure, and more. But maybe your stressors aren't quite so insurmountable. Maybe the things that stress you out are public speaking or long hours or recalcitrant children. If you could reduce the amount of stress in your life, then you would stop producing the cortisol and glutamate and adrenaline and other stress-related chemicals when you don't actually need them, thereby reducing your risk of heart disease and hypertension. But you don't have to stop there. If you can *increase* your happiness, and not just *decrease* your stress, you will actually get healthier. Happiness is good for you!

People who are engaged with life, enthusiastic, hopeful, or content are far less likely to suffer from coronary disease, even if they smoke and don't exercise. Happiness is shown to boost your immune system and lower accounts of aches and pains. Not just in the short term, either. Happy people are simply less likely to suffer from long term chronic illnesses. Even among the elderly, happy people are less frail and more vigorous, and happy people simply live longer, by as much as 7-10 years[91]. In truth, researchers are still untangling cause and effect here. In whole or in part, it may be that good health causes happiness and not the other way around. Personally, I suspect it's another virtuous cycle.

[91] I'm not going to dive deep into the relationship between happiness and good health. If you're interested, here's a good place to start: https://greatergood.berkeley.edu/article/item/six_ways_happiness_is_good_for_your_health.

Keep in mind that happiness is not a switch. You can exercise all you want, and you'd be better for it, but if you can't pay the bills, you're still going to be under extreme levels of stress. There's no magic here. The sources of happiness, true happiness, are large and multifarious. Safety in one's person is vitally important. Financial security is likely a close second. I imagine most people would add job satisfaction, a loving family, intellectual fulfillment, good friends, and many more. It just happens that getting and staying healthy is an important component of being happy, as well.

You may contend that soda and cheesecake also make you happy. To which I will say, absolutely they do. But please don't mistake momentary bursts of pleasure for true happiness. Pleasure is fleeting, happiness lasts. Don't take what should be an occasional and enjoyable treat and turn it into a necessity. Candies and cakes, unlike nicotine, are not something you need to quit entirely; you just need to be aware of how much you're consuming. If you have a small glass of soda every once in a while, the world won't end. If cheesecake brings you pleasure, then by all means have a slice of cheesecake on occasion, you've earned it. Similarly, if you want to take a week off from exercising just to reset, then do so. Better still, schedule your bye-weeks to coincide with your vacations. And while you're on vacation, eat whatever the hell you want. Being healthy does not mean a life of asceticism. Go ahead and continue to do things that bring you pleasure, just add "eat well and exercise regularly" to the list.

By the way, if you think cheesecake is good, I've got another treat for you. When you have a healthy diet and exercise regimen, suddenly *all* food is cheesecake-good. They say that when you quit smoking, food tastes better. Something to do with your nose being more sensitive to subtle aromas. Maybe that's true for some people, but I didn't notice it. What I found makes food taste better is hunger, and exercise makes you hungry. A moderately active male of my age, height, and weight burns around 2,600 calories just getting through his day. If he's not trying to lose weight, all those calories need to be replaced. Add a three-mile jog in the morning and an hour of strength training in the evening, and now he needs to consume 3,100 calories to stay at replacement level. If he's trying to build muscle, add 200 more calories on top of that. Now imagine it's a crisp autumn Saturday. Our calorie-starved protagonist gets up early, puts on his running clothes, and goes for a three-mile run in the park. Do you know what tastes better after that run than a medium rare New York strip with two eggs sunny side up, steamed broccoli, rice, and buttered toast (around 1,000 calories)? Nothing! Nothing at all tastes better than that. Certainly not cheesecake. Not even my sister Mary's cheesecake, and my sister Mary's cheesecake is awesome.

I know this because I've had that breakfast under those conditions, and it made me deliriously happy. The steak, broccoli, and rice were all left over from the night before, so really, I only had to make eggs and toast. It was delicious, filling, and easy. As are even the most workaday meals I have. The PBJ (natural peanut butter, whole grain bread) and pint of milk for lunch (900 calories[92]). The spinach salad with grilled chicken, feta cheese, toasted pine nuts, dried strawberries, roma tomatoes, red onion, and a vinaigrette dressing, with a buttered baguette and some wine for dinner (1,000 calories more[93]). Those taste as good as a cheesecake any day. And if my calories estimates are correct, at 2,900 calories, there's still room for some cheesecake, so go ahead and have a slice. Just don't make a habit of it.

It's not just the actual taste of the food. I also get a charge from knowing that I'm eating well. If I'm being honest, I'm actually a bit smug, but I feel good about preparing and eating healthy, delicious meals. I also feel like a better father when I set a grilled chicken salad in front of the children instead of a plate of spaghetti with sauce from a jar. Although, around the dinner table of late, the boys have expressed other opinions about my suitability as a parent.

Cigarettes used to make me happy. It was largely the false happiness of addiction, of course, but there was also the caffeine-like stimulant, the alcohol-like sedative, and the Bogart-like cool. Unlike with added sugar, however, you cannot reward yourself with an occasional cigarette; the risk of relapse is too great. If you're a smoker, you need to give up smoking entirely. Forever. If this scares you, then you should know that besides the obvious health and financial benefits of not smoking, there are some other unexpected upsides to quitting. For one thing, airports and flying are no longer a source of dread (at least when it comes to satisfying a nicotine addiction). Similarly, on nights out with your SO, he or she will appreciate that you don't disappear for three minutes right after the appetizers are finished. And there's no need to annoy the entire family by pulling over every hour during long car rides.

You may not ever have smoked, but if you're reading this book, especially if you're American, then there's a better than even chance that you're out of shape. Here's some non-startling information: being fit also makes you happier. There are lots of words for this feeling. In order from most to least generous, I'm thinking of self-esteem, pride, and vanity. Where I am on this spectrum varies from day to day and from moment to moment, but the fact remains that it makes me happy when people remark on my appearance, from the simple "You've lost weight" to the quips about being jacked. I'm proud that I actually did quit

[92] Milk: 300 calories, bread: 150, peanut butter: 300 (I slather it on), jelly: 150

[93] Call it 400 calories for the salad, 400 for the wine (2 glasses), and 200 more for the bread and butter.

smoking, and then took up running. I like wearing figure-flattering clothes, even if they're just jeans and a tee-shirt. I'm proud that I can climb a mountain. I like running into people I haven't seen in a while. I like how I look in a mirror. I ran a ten-mile race! Heck, I even look forward to going to the beach.

I joked earlier about the cult of CrossFit, but the truth is I was flattered that people capable of doing CrossFit ever considered inviting me to join. To no small degree, my brain hasn't quite caught up with my body, and it still sometimes thinks of me as the feeble, overweight, smoker. I'm always getting tripped up, if only briefly, when athletic people treat me as one of their tribe. When the jogger on the other side of the street shoots me a quick wave, I wonder if they're waving at me. When the guy at the pool asks what gym I go to, I wonder why he's asking me. But then I remember, and a flush of pride comes over me. I like that.

Another source of pride, something a little further away from narcissism, is the "PB." That's how hard-core athletes refer to their "personal best." For some people it's always necessary to be chasing a PB, whether that be most marathons in a year or how much they can bench. That's understandable, but I'm not hard core, I will let things slide and even reverse for a while if I want to. That said, I do like to see progress. Like everyone else, I like the sense of accomplishment that comes from reaching a goal (that's the dopamine again). I like it when I finally complete three sets at a given weight and I get to load up next week. It makes me happy when the *Runkeeper* app points out that I just did my fastest 5K. I'm over the moon when I summit a 4,000+ foot mountain. And each of these accomplishments is sweetened by the fact that I couldn't — indeed, wouldn't — do any of these things just a few months ago. In short, I get a kick out of being able to say to myself, "I did that. What next?"

And, hey, the aches and pains are gone! That was the original goal, and it makes me happy that I can now sneeze without fear. I'm also happy that so many other recently difficult tasks are now routine again: taking out the trash, moving furniture around, chasing after things. Not only that, but according to Denise, I've stopped snoring. That was a pleasant surprise for both of us. As nice as all this is for me, for others the simple pleasure of living without pain or discomfort might be even more pronounced. I was overweight, but not obese. Nor was I managing a chronic condition. If you are, I can only imagine that getting to a healthy weight and enjoying the food you eat and your time in the gym is all the more gratifying.

Notice I said "healthy weight." As briefly noted in Chapter 3, healthy doesn't always mean slender. You can be fat and fit (although it's a lot harder to be obese and fit). Body weight is highly heritable. I, for instance, am a skinny person. My father was skinny, as was his father before him. It's only in the last 15 years that

I tacked on a few unwanted pounds. Being mentally and genetically skinny likely made it easier for me to lose that excess weight and keep it off. The same may not hold for you. So long as you eat whole foods, exercise regularly, and don't smoke, you're going to be great. Don't worry about your weight too much, just get healthy, the rest will follow.

Love is the Drug

Beyond the simple joys expressed above, there's one more wholly unanticipated change brought about by my adoption of a healthy lifestyle: I love my wife more.

All of the other frisson-inducing changes — the sense of accomplishment, the self-esteem, the food, even the chemical cocktails being mixed up in my brain — they're all small-ball stuff compared to simply being more alive. That I love my wife more than I did a year ago is just the last in a huge cascade of consequences that stem from this simple fact: I am more engaged in life than I ever was before. Mine and hers.

At its simplest, it is now the case that Denise and I inhabit much the same world. Sure, she'll never read *The Silmarillion*, but I'll never read *Bleak House.* Our differing tastes in literature, however, is just one of those endearing things that make people interesting and unique. What I mean about inhabiting the same world is that now, when Denise talks about an especially tough yoga class, I'm interested in hearing what made it so hard and how the buff guy in front of her almost cried. And when I say I'm going for a hike, she'll help me get my gear together and listen to me gush on about it when I get home. Although, it's more likely that she will simply join me. We're on exactly the same page when it comes to what we eat: keeping it healthy, eyeballing those calories, avoiding excess sugar. We tell each other our weight, and we're interested in how the other is faring.

More importantly, we frequently run together or go to the gym together or walk the dog together. These are not just 20-minute affairs; it's a significant chunk of time that I now spend with my wife deciding on where and how far to run, then getting dressed, getting out, doing the run, and getting home — that's not even including the weeks we spent training for the 10-miler. On top of that, afterwards, we have all these pleasure chemicals sloshing around in our skulls, as well as a shared experience and a sense of accomplishment, and we feel good about ourselves and each other. Honestly, it's not that different from sex, which, when you think about it, is pretty fucking cool.

I need to take a moment to better describe Denise's metamorphosis. As you know, I have not been alone in making lifestyle changes. Compared to me, of course, Denise is a veritable saint. She was always physically active, though her level of activity would ebb and flow; she might have two drinks in a single night, but if so, that's it for the week; and she's never smoked a day in her life. She was, however, overweight, every bit as much as I was. When I hit 210, Denise was 30 pounds behind. As with me, that was officially 20 pounds overweight. Unofficially, though, that was 50 pounds more than she wanted hanging on her five-foot seven frame.

While Denise did not set the reinvigoration dial to eleven, she didn't need to. She only needed to lose weight, and she did. Today she weighs in at 140 pounds and is on the fence about losing 10 pounds more. It's not just a change in diet (although strict adherence to the ketogenic diet enabled her to lose 30 pounds in three months); like me, Denise is spending much more time exercising. She does yoga for strength training, and running (warm weather) or cross-country skiing (cold weather) for cardio.

It's not an accident that Denise's weight loss and renewed interest in exercise have coincided with mine. We did this together (which itself has brought us closer). Denise has been yo-yo dieting and sporadically exercising since I met her. But now, with me there to *finally* be supportive and, I dare say, to emulate, she's pushing herself farther than she ever has before. It shouldn't need repeating, but the opposite is also true. There's no way in hell I could have succeeded in getting fit without Denise kicking things off and spurring me on. I think just this level of mutual support during such dramatic change has helped us get better at supporting each other in general.

And now she feels better physically and mentally than she has at any time since I met her. Her energy is boundless. Even as I write this, she is out for a run. I got my three miles in this morning, while she drove into Cambridge to meet a client. That's okay, we don't always need to exercise together. But all of this is exemplary of the new-found vitality in our lives. And this vitality is starting to permeate everything we do: our conversations, our meals, our friendships. Maybe it's cliché, but we feel alive. That's a level of happiness that dwarfs pride in appearance or the pleasure of a good meal any day, and it's a direct result of being healthy.

So, yes, I love my wife more today than I did a year ago. Are we allowed to admit such things or do we have to pretend to some romantic ideal of love that doesn't alter when it alteration finds? Does that hold, even if the change is for the better? The fact is, having shared interests and common experiences has brought Denise and I closer. Having this happen at the same time we are getting increasingly healthy and feeling good about ourselves is icing on the cake, so to speak. Actually, not "at the same time," but rather "because of." Because we're getting healthy, we have more opportunity to be together doing something we both enjoy. Because we're getting healthy, we feel better physically and emotionally. Because we're getting healthy, we are more supportive and in love.

Though, I suppose it could also be the oxytocin.

What galls me is that I could have felt this way my entire adult life! That I actively chose not to out of laziness and ignorance is going to gnaw at me forever. Happiness is not always twinned with regret, but in this case it is, and it kind of sucks. I should have quit smoking decades ago. Better still, I should never have started. But doing stupid things doesn't bother me; it's the continuing to do stupid things for years part that I regret.

To be clear, it's not the euphoria of exercise or the joys of healthy eating that I regret missing out on. Those would have been nice, of course, but they are more of an added bonus than a primary source of happiness. What really makes me remorseful is knowing how much of life I simply let pass me by. For along with the vitality of good health comes a desire to say "yes" to what life has to offer. When you realize this, it becomes painfully clear how many times you've said "no." As discussed in Chapter 1, I said a lot of "no" over the years, explicitly or not, just because I didn't feel like exerting myself or leaving my comfort zone. To be even more clear, while I miss all the runs I never ran and all the tennis I never played, I am truly rueful that in missing out on these things, I really missed out on the opportunity to do them with friends and family.

But just as the little aches and pains are gone, so too is my desire to say no: to my kids, to my wife, and to myself. Getting healthy is paying dividends far in excess of what I expected. Yes, I can tie my shoes again, but I can also play basketball with the kids, and run with my wife, and laugh with my friends. Hell, not only can I run the three blocks from the downtown playground to my house, I can run all the way to the hospital — and back, several times. All is not perfect, sadly. I kind of screwed up the timing, and Henry has reached that age where he's starting to pull away from his parents. That said, he's not gone yet.

So H-Bomb (we call Henry "H-Bomb" or just "H"), listen up. About a year ago, before all this getting healthy stuff even started, I stumbled across the baseball glove I bought when you were in Little League (because, of course, I didn't already own one). Since that glove never got a lot of use, I dug out the glove oil I bought at the same time and worked it into the leather. I found a baseball and some rubber bands, and wrapped the glove up tight. It's in the chest by the front door where we keep all the other sports equipment we don't use. How about I dig it out, and you call — or text or snap or insta-whatever —

Tommy and Tyler, and we all head over to Cashman Park and toss a ball around the horn a few times? I'll even chase down some flies if you want to.

And Brosie (we call Ambrose "Brosie" or "The Brose," but strangely "A-Bomb" never stuck), I know baseball's not your thing, and neither is football or soccer or even frisbee. But if you can find a friend who's willing to DM, I would love to get in a Dungeons & Dragons game with you. We can even use the dining room table, and I'll supply the snacks. How many parents would be cool with that, huh? My mother always thought we were literally summoning Astaroth. Which, when you think about it, is pretty stupid. After all, Astaroth is way too powerful to play against a 4th level magic user. The game would be way too unbalanced. (Yes, of all the things that require no amount of fitness whatsoever, D&D is high on the list. But feh! I still want to play.)

And dearest Nisey (I call Denise "Nisey" sometimes; rhymes with greasy), I am genuinely sorry that I have squandered so much of the last 20 years, and I wail my dear time's waste. I apologize for not being there emotionally for you and the boys, and for almost destroying a marriage that I now know could not only have been happy but exceedingly so. I want to assure you that things are different now. The last thing I want to do at this point is passively observe life. I want to run, hike, eat (well), and drink (a little too much). I want to explore, engage, and try new things. And I want to do all these things with you. While the last few years could have been better, by many measures they were pretty good. We have two wonderful children, a nice home in a nice town, good careers, and financial security. While I wish I hadn't waited so long to fully engage with life and love, I'm here now. The odds are we just bought ourselves a few more years on this rock, and I promise, those years will be our best.

In My Time of Dying

Which brings us to the little matter of death.

Back in May of last year, Denise and I had a very simple plan. She and I would participate in a brief exercise course in the hopes of putting an end to some of life's many indignities. I did not plan to take up aerobic exercise, nor did I plan to take up strength training. I did not plan to quit smoking, change my diet, lose weight, or in any other way to become healthy. But here I am. If not healthy, at least healthier, and more so every day. Being generally and increasingly fit means more than being able to shovel snow without fear. It very likely means I will live a longer and more fulfilling life.

Upon reaching their 50th year, everyone becomes a bit of an actuary, with their thoughts straying more and more toward one particular bit of statistical reasoning: How long am I likely to live? This math is informed by your current age, the average age of death for your sex and socioeconomic background, your family history, and your lifestyle. In my case, the math wasn't working out so great. But being who I was, I treated lightly the possibility of shuffling off this mortal coil before reaching 65. It was a form of gallows humor, I suppose. Denise did it too. She and her girlfriends would repeatedly jest about all the fun they'd have together once unburdened by husbands (who weren't running 5Ks either).

That I was once comfortable joking about having only a few more years to spend with my family is something I now find galling. Even if I had been granted fifteen years more of life, I would be only 68 when I died. If Henry and Ambrose refused to listen to me and married young, then I might get to attend a wedding or two, maybe even hold a grandchild in my lap. But more likely, fifteen years wouldn't mean any of that. No weddings, no grandchildren, no chance to watch my boys change the world. Hell, The Brose will likely still be living in my house.

And what of Denise? Maybe she and her girlfriends would have been as happy as they say. But damn, leaving her a widow before 70 for no other reason than because I was too fat and lazy to lift my butt out of chair! Exactly how fucking selfish do you have to be? And when I was gone, would she have any reason to mourn?

It wouldn't just be the number of days between then and my time of dying, but the quality of those days as well. In my case, not only would I be battling my lifestyle and suffering through one or more of lung cancer, emphysema, diabetes, heart disease, atherosclerosis, hypertension, etcetera, but I would also be wrestling with a, shall we say, suspect set of genes.

◆ ◆ ◆

My father, Ambrose Lacey, was born in 1928 on a storm-battered rock off the western coast of Ireland, where he was promptly abandoned by his parents, who skipped to America. An only child, he was placed in his aunt's care, who raised him until he was nine years old. At that tender age, he was put, unaccompanied, on a steamer to New York, where his father was working as a longshoreman and his mother as a seamstress. Back then, a longshoreman's salary could get you a place in Park Slope, Brooklyn, and for the next seven years he shared a residence there with his parents, though they shared little else. In 1944, lying about his age, he joined the U.S. Navy, and shortly thereafter was shipped off to the Pacific theater, where he saw action as a machine gunner on a landing craft. After a stint in occupied Japan, he returned to New York, got his undergraduate degree on the GI bill, and in 1958, at 30 years old, met and married my mother.

Despite his stony upbringing, he was a good man. He was a loving husband and (in an Irish sort of way) a caring father, active in his church and respected by the community. He spent the next 35 years working for AT&T, raising a family, and maintaining a modest house that at one time had to sleep ten people, because he had also taken in his (cantankerous) father-in-law and his own (drunken) father after their wives had passed away. I'm told he used to smoke, but he quit before I can remember. I'm also told he drank to excess when he was younger, but for as long as I can recall it was one can of Rheingold a night (until Rheingold went out of business and he switched to Budweiser). When the kids were older and the house was in order, he spent his Saturdays fishing on the Dorothy B. out of Sheepshead Bay and his Sundays as a sexton for the church. As a member of the Holy Name Society, he helped arrange community events in the Catholic school gymnasium, such as the St. Patrick's Day dance and Golden Gloves boxing matches, where my sister Eileen did coat check and I slung pitchers of beer. While he didn't exercise the way people today exercise, he was a robust and active man. Until he suddenly wasn't.

In 1990, complaining of chest pains, he was diagnosed with coronary artery disease (clogged arteries). By all accounts, the subsequent triple bypass was a success. He retired from the phone company and presumably planned on a long, if bacon-free, retirement. But soon after the surgery, the first signs of dementia became evident: getting lost in neighborhoods he knew well, mixing up my sisters' names, and so on. By this time, I was living in Chicago, so while the rest of my family had to watch the daily progression of his disease, I only caught snapshots of it during trips home for holidays and family events, which only served to make the decline seem more dramatic. On Thanksgiving I witnessed sudden bursts of anger when he realized that his skills and faculties were slipping away. At Christmas he would tell the same tired joke over and over. At the next christening, he would not recognize his grandkids. By the time of my wedding, eight years later, he didn't even recognize me.

Ultimately, he was becoming a danger to himself and my mother, and my sisters had to put him in a long-term care facility. I visited him there just once, I'm ashamed to say. Of course, he had no recollection of who I was. His mind was so far gone we couldn't even converse as strangers would. Instead, I pretended he could understand me and told him how my life was going, and he would reply with nonsensical phrases and the names of people I did not know. Toward the end of my disgracefully brief visit, I walked with him to the receiving area and signaled a nurse. Pushing the button for the elevator, I gave him a hug and a kiss goodbye, and told him I'd be back soon. The elevator doors opened and I stepped inside, when my father shouted, "Wait!" I turned, and his eyes seemed suddenly lucid as he looked pleadingly into mine. "Where am I?" he asked, confusion in his voice. My breath caught in my throat, the nurse took him by the arm, the elevator doors closed, and I fell back against the elevator wall, rocked to the core.

◆ ◆ ◆

My mother, Anna Fitzpatrick, was born in Brooklyn in 1934. (If you knew the name of my first pet, you'd be able to empty my bank account by now.) She was raised in St. Teresa's parish, which is how the Irish referred to the neighborhood everyone else knew as Crown Heights. One of six kids, she had what I understand to be a normal Depression-era childhood, ultimately finishing high school and getting one of the few jobs open to her, telephone operator. She met my father at a church dance and married him at the age of 24. Then she spent the next ten years spitting out and raising six kids of her own.

Sometime in the mid-70s, however, she took ill — and stayed that way. From the time I was 10 or 11, my predominant memory of my mother was her lying in bed, too sick to function. This was not constant, it would come and go in bouts, but the bouts would last weeks. She was so incapacitated during these spells that she kept a large bell by the side of the bed, which she would ring to get the attention of one of us kids when she needed something. The doctors diagnosed her with and — likely worse — treated her for ileitis and colitis, which presents as inflammation and ulceration of the bowels. Unfortunately, what she really had was a hypothyroid, which can present in the same way.

It wasn't until around 1985 that some doctor finally twigged to the hypothyroidism and began the appropriate treatment. She was a woman reborn! The weight she had gained from constant convalescence and a thyroid-less metabolism melted away. She lost the housecoats and the curlers and acquired a new wardrobe. She got a job! A job working for Kelly Services — *for*, mind you, not *as*. She placed Kelly Girls, which, believe it or not, was a term they still used in the mid-80s.

For a few years there, everything was coming up Lacey. But then my father had his surgery and my mother quit her job to tend to him. Sadly, my father's slow deterioration sucked the life out of the poor woman. She had built her world around that man, and as he slipped away she started putting on weight — lots of it. She began to spend most of her time in bed again, finding it increasingly hard to get around. She never did quit smoking, but developed a candy habit to go with it, so there's that. By the time of my wedding, when she was just 64 years old, she was largely confined to a wheelchair. At family gatherings, if stairs were involved, the men of the family would link their arms to form a sedan chair of sorts and carry her in and out of the house. Ultimately, she acquired all the diseases associated with poor nutrition, obesity, and lack of exercise: diabetes, congestive heart failure, skin ulcers, and more.

By 2009, when our mother was no longer able to live alone, my sister Mary took her in. By 2012, when that became no longer tenable, my mother was also entered into a long-term care facility. She lingered there for four years, unable to get out of bed, her organs failing, the pain nearly constant. Finally, in 2015, she passed away, having spent the previous decade and more existing but not living, pissing into a tube, and unable even to roll over.

◈ ◆ ◈

That's the genetic hand I (and my siblings) have been dealt. From Dad I got a full head of hair, a wiry frame, an increased risk of heart disease, and possibly an increased risk of dementia. From Mom, I got a slightly increased risk of hypothyroidism and, for lack of a better term, depression. Up until recently, I never considered that she may have suffered from depression, but in retrospect I can't think of any other rationale for the way she simply gave up on life. In short, both of my parents spent their last decade or two living but not alive, and that fate could have easily been mine.

Had I not quit smoking, lost weight, changed my diet, and started a cardio program, it's extremely likely that I would have had some form of heart failure in the next ten years or so. In addition, all of those lifestyle choices, along with the alcohol consumption, are implicated in the onset and progression of dementia. In all likelihood, I have now staved that off for a while. There's also a strong causal link between heart health and brain health, which may buy me a few more years on top of that. All in all, while I can't say that heart failure or Alzheimer's aren't in my future, I have greatly reduced my risk of contracting either[94]. In fact, I've reduced my risk to well below average for an American.

As for my mother's genetic contribution, maybe I'm in the clear. She was chronically ill for quite some time, and that can't have helped. Whereas I rarely get sick at all. If I've ever missed a day of work due to illness, it was well over a decade ago. In addition, my mother inarguably had some mental health issues. While I regret those recent years of being more distant from my family and friends

---

[94] This is especially so since Alzheimer's disease, sometimes dubbed "type 3 diabetes," is linked so strongly to insulin resistance.

than I should have been, I just scanned the Internet, and I'm certain I don't suffer from any form of clinical depression. I just got lazy, or, more honestly, addicted.

That said, there were a lot of people in those long-term care facilities, and not a single one of them seemed even remotely vital. They were broken and alone, barely able to move and interact. Unable to fix themselves a sandwich, go for a walk, or wipe their own ass. I suppose everyone thinks they won't end up like that, and that includes the people who were there. They had childhoods, got jobs, bought houses, raised families, and now they wither away under fluorescent lights and the doleful, pain-induced keening of their neighbors.

Well, fuck that noise. If there's one thing I hope I have made clear by now, it's that a healthy lifestyle, i.e., aerobic and resistance exercise, a wholesome diet, and not smoking, is the next best thing to magic when it comes to a long and vibrant life. It's like having a cleric NPC in your party who hangs back in every combat situation just casting healing spells. Go for a run, roll 2d8. Eat a salad, roll 2d4. Other factors, notably moderate drinking and a strong IRL social network, are also important, but I'm still working on those.

But I didn't become fit so that I can better enjoy my golden years. As is the underlying thread of this book, I didn't intend to become fit at all. But now that I am healthy and can actually see a better future for myself, you can be damned sure I'm going to stay that way. I know that forty years of abuse and neglect has done real, maybe irreparable, harm to my body, but it doesn't feel that way. Everything seems to be in full working order and I feel great. My job now is to keep to the maintenance schedule.

So, yeah, I am going to rage against the dying of the light. I am going to burn and rave and blaze like a meteor. I refuse to spend the last years of my life lucid but trapped in a failing body of my own making. Indeed, I want the distance between my body betraying me and my ultimate demise to be as short as possible. Seconds would be good. And I will happily — HAPPILY — forego the cigarettes and donuts and television, to keep far from the compass of Time's bending sickle. I'm just pissed that I waited so long.

Long May You Run

By convention, this is the part of the book where I bestow upon you the secret behind my newly discovered enlightenment. This is where I distill the essence of my unique and formidable journey to insight and understanding, so that you, my soon-to-be disciple, can apply these learnings to your life and share in an exalted future. All I need to do is list ten — no, scratch that, three — three simple steps to long life, happiness, and loving relationships. Steps no one has ever thought of before. Steps that, if I play my cards right, come at a low, low price that you would be a fool to pass up. It would be nice, lucrative too, but I'm sorry, no such denouement is at hand.

I haven't done anything all that special. Yes, I lost 30 pounds, but lots of people have lost that amount of weight and more — much more. It's true that I changed my diet to limit processed and sugary foods, but a whole host of others have done that too. Sure, I quit smoking, but so have millions of Americans. I also gave up a lifetime of indolence for an expansive exercise regimen of interval training, endurance training, and strength training, but that's hardly special. In truth, I *do* think it's considerably less common to have done all these things in such a small amount of time, especially at 53 years old, but "less common" is not the same thing as "unique." As I alluded to in the preface, the skies have not opened up to me and there is no revelatory moment for me to share. I have not followed the Eightfold Path, the scales have not fallen from my eyes, and I have not hung from the branches of Yggdrasil. I am an ordinary man, living an ordinary life.

And I am deeply and profoundly changed.

But if you insist:

- Eat real food
- Exercise
- Don't smoke

I can add one more thing, something I've been alluding to throughout this book: don't tackle everything at once. That's too ambitious, a recipe for failure and self-recrimination. Don't set out to change your life; instead, have only the goal of being a little healthier every day. Don't plan anything, don't set a schedule, just keep on surprising yourself. When you're ready, you'll have to force yourself to take that first step. That's the hardest one. If you're lucky, you'll have a partner to help you. If not, you'll need to find it within yourself to make it happen. It doesn't have to be anything as daunting as boot camp. Maybe just cut out the sugary drinks. Sometime after that, consider reducing the amount of refined flour in your diet. Then go for a walk; it's only 30 minutes five times a week. You can do that. And if you never do any more cardio than that, that's just fine. If you take up running, however, welcome to the club. Try to cook more. Go for a bike ride. Call an old friend. Do some pushups. Enjoy the outdoors. Rewire your brain. Play with the kids. Oh, and get plenty of sleep.

I'll see you out there.

Acknowledgements

Many people contributed to the publication of *Scientifically Sound.*

The cover art is from a public domain image available on Wikipedia. It depicts the crystal structure of a myosin V motor protein. https://commons.wikimedia.org/wiki/File:Myosin_V.png

Therese Arkenberg reviewed multiple drafts of this book, carefully editing each sentence and, to the book's great benefit, suggesting multiple structural improvements. It's safe to say that *Scientifically Sound* would be unreadable without Therese's invaluable contributions. Thank you.

Marie McGovern, PhD and Brigid Goldman, PhD both volunteered their precious time to review this book for scientific accuracy, and I couldn't be more grateful. Their efforts have vastly improved this work and saved me from untold embarrassment. I am deeply appreciative of the time they spent correcting my many errors and providing useful advice and information. Needless to say, any remaining errors are entirely my own.

My thanks also go out to Brian McGovern, who made me realize that I should not be writing a book about me, but rather about you. And also, that I start too many sentences with the word "And."

Most importantly, I would like to thank my wife, friend, and running partner, Denise Lacey. Denise played an oversized role on my journey to good health. For that, as for so many things, I will be forever in her debt.

Made in United States
Cleveland, OH
04 November 2026

47650807R00099